Authors and Reviewers

Tricia Salerno
Jenny Kempe
Bill Jackson
Allison Coates

Workbook

DIMENSIONS MATH 6B

Published by Singapore Math Inc.

19535 SW 129th Avenue
Tualatin, OR 97062
www.singaporemath.com

Dimensions Math® Workbook 6B
ISBN 978-1-947226-43-2

First published 2018
Reprinted 2019 (twice), 2020 (twice), 2021

Copyright © 2017 by Singapore Math Inc.

All rights reserved. This book or any portion thereof may not be reproduced or used in any manner whatsoever without the express written permission of the publisher.

Printed in China

Contents

Chapter 8
Algebraic Expressions

8.1 Writing and Evaluating Algebraic Expressions
   A. Use of Letters 1
   B. Evaluating Algebraic Expressions 6
   C. Word Problems 9
8.2 Simplifying Algebraic Expressions 14

Chapter 9
Equations and Inequalities

9.1 Equations
   A. Algebraic Equations 19
   B. Balancing Equations 22
9.2 Inequalities
   A. Algebraic Inequalities 34
   B. Graphing Inequalities Using a Number Line 35

Chapter 10
Coordinates and Graphs

10.1 The Coordinate Plane 40
10.2 Distance between Coordinate Pairs 44
10.3 Changes in Quantities
   A. Independent and Dependent Variables 57
   B. Representing Relationship between Variables 58
   C. Observing Relations between Variables with Graphs 61

Chapter 11
Area of Plane Figures

11.1 Area of Rectangles and Parallelograms ... 72
11.2 Area of Triangles
   A. Finding Area of a Triangle ... 78
   B. Areas Involving Parallelograms and Triangles ... 84
11.3 Area of Trapezoids ... 88

Chapter 12
Volume and Surface Area of Solids

12.1 Volume of Rectangular Prisms
   A. Cubes and Cuboids ... 93
   B. Volume of Liquids ... 101
12.2 Surface Area of Prisms
   A. Surface Area of Rectangular Prisms ... 110
   B. Surface Area of Triangular Prisms ... 119
   C. Metric Conversions and Volume ... 122

Chapter 13
Displaying and Comparing Data

13.1 Statistical Variability
   AB. Statistical Questions & Measures of Center ... 126
13.2 Displaying Numerical Data
   A. Dot Plots ... 134
   B. Histograms ... 137
13.3 Measures of Variability and Box Plots
   A. Range ... 145
   B. Mean Absolute Deviation ... 147
   C. Interquartile Range ... 151
   D. Box Plot ... 154

Answer Key ... 156

8 Algebraic Expressions

8.1A Use of Letters

Basics

1. For each of the following statements, write an algebraic expression.

    (a) 5 less than s _____

    (b) k less than $15\frac{1}{2}$ _____

    (c) 2.25 more than h _____

    (d) m more than 100 _____

2. (a) The product of $\frac{5}{6}$ and c _____

    (b) The product of e and 8.2 _____

    (c) The product of 1,000 and r _____

    (d) The product of m and c^2 _____

    (e) The quotient when t is divided by 12 _____

Algebraic Expressions — 8.1 Writing and Evaluating Algebraic Expressions

3. Explain why $\frac{s}{10}$ and $\frac{1}{10}s$ are equivalent.

4. Using fractions, write the algebraic expressions for each of the following statements in two ways.

   (a) The quotient when $5k$ is divided by 3 _____ _____

   (b) The quotient when $4c$ is divided by 2 _____ _____

5. Write an algebraic expression for each of the following statements.

   (a) 5 less than the product of h and 3 _____

   (b) 10 more than the quotient when a is divided by 5 _____

   (c) $\frac{2}{5}$ plus the product of $\frac{3}{10}$ and k _____

   (d) $\frac{1}{2}$ minus the quotient when $3t$ is divided by 5 _____

   (e) c more than $7a$ _____

   (f) $10p$ more than $7n$ _____

Practice

6. Emiliano was the first student to finish a quiz. He finished this quiz in k minutes. For the same quiz, Dexter took 8 minutes. Write an expression that expresses the difference in time it took the two boys to finish the quiz.

7. Nina earned p points on a math quiz. Express your answers to the following questions in terms of p.

  (a) Kale earned twice as many points on the math quiz as Nina. How many points did Kale earn? _____

  (b) Salim earned $\frac{2}{3}$ as many points on the quiz as Nina. How many points did Salim earn? _____

  (c) Papina earned 2 fewer points on the quiz as Nina. How many points did Papina earn? _____

8. Aki has p fewer apps on her phone than Fadiya. Fadiya has 23 apps. How many apps does Aki have?

9. A blouse costs *b* dollars and a pair of pants costs *p* dollars. How much will 3 blouses and 2 pairs of pants cost?

10. Ella bought *c* cookies. She ate 2 of them, and then split equally among the 3 friends. How many cookies did each friend get?

11. A serving of strawberries weighs *s* grams. A serving of cherries weighs *c* grams. How much do 3 servings of strawberries and 2 servings of cherries weigh?

Challenge

12. The length of a rectangle, l cm, is equal to its width, w cm, squared. Express the area of the rectangle in terms of w.

13. Santiago collects sports cards. The ratio of his baseball cards to his hockey cards is 4 : 3. He has $32b$ baseball cards. He has n more basketball cards than hockey cards. How many basketball cards does he have? Express your answer in terms of b.

8 Algebraic Expressions
8.1B Evaluating Algebraic Expressions

Basics

14. Evaluate the following expressions when $y = 3$.

    (a) $y + 10$ _____   (b) $5y$ _____

    (c) y^4 _____   (d) $\frac{4y}{6}$ _____

15. Evaluate the following expressions when $n = 10$.

    (a) $\frac{3}{5}n$ _____   (b) $1.2n$ _____

    (c) $\frac{5}{3}n$ _____   (d) $0.07n$ _____

16. Evaluate the following expressions when $h = 6$.

    (a) $7 + 3h$ _____

    (b) $9h - 22$ _____

    (c) $(18 - 2h) \div 4$ _____

    (d) $18 - \frac{2h}{4}$ _____

    (e) $\frac{1}{3}h + 15$ _____

Practice

17. Evaluate the following expressions when $m = \frac{1}{2}$.

(a) $6m - 2$ _____

(b) $5 + 4m$ _____

(c) $\frac{9}{14} - 4m \div 7$ _____

(d) $\frac{9}{14} - \frac{4m}{7}$ _____

(e) $\frac{1}{2}m - \frac{1}{4}$ _____

18. Evaluate the following expressions when $s = \frac{3}{4}$.

(a) $\frac{s}{6}$ _____

(b) $\frac{2s}{8}$ _____

(c) $\frac{s}{12} + 9$ _____

(d) $\frac{8s}{5} - \frac{5}{6}$ _____

19. Evaluate the following expressions when $b = 0.45$.

(a) $\frac{3b}{5}$ _____

(b) b^2 _____

Algebraic Expressions — 8.1 Writing and Evaluating Algebraic Expressions

> **Challenge**

20. Evaluate the following expressions when $a = \frac{2}{3}$, $b = \frac{1}{2}$, and $c = \frac{3}{4}$.

(a) $\frac{a}{b} \times 8c$

(b) $3a + \frac{b}{8} - c^2$

8 Algebraic Expressions

8.1C Word Problems

Basics

21. Identical flower arrangements have 12 flowers in each arrangement.

    (a) Complete the table.

Number of Arrangements	Total Number of Flowers
1	
2	
3	
4	
5	

    (b) If the florist has f flowers, how many possible arrangements could be made?

    (c) How many flower arrangements can be made with 180 flowers?

Practice

22. A bean plant grown in the shade is 3 cm shorter than a bean plant grown in the sun.

(a) The sun-grown plant is b cm long. Express the height of the shade-grown bean plant in terms of the sun-grown plant.

(b) What is the height of the shade-grown bean plant if the sun-grown bean plant is 32 cm tall?

23. The Dang family owns 2 desks. One is a valuable antique, and the other was made recently. The antique desk was made 75 years before the modern desk.

(a) Given that the modern desk was made d years ago, express the age of the antique desk in terms of d.

(b) How old is the antique desk if the modern desk was made 4 years ago?

24. A rectangular garden has a length of g meters, and a width of 8.6 meters.

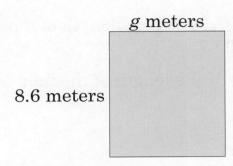

(a) Write an algebraic expression to find the area of the garden.

(b) Express the area of the garden in square meters in terms of g.

(c) Find the area of the rectangular garden when $g = 7.8$.

25. In a pet shop, there are 2.5 times as many dogs as cats.

(a) If the number of cats is c, express the number of dogs.

(b) If there are 4 cats in the pet shop, how many dogs are there?

26. The ratio of the length of a rectangular field to the width of the field is 3 : 7.

(a) If the length of the field is l, express the width in terms of l.

(b) If the width of the field is 28 meters, what is the length of the field?

> **Challenge**

27. There are 3 times as many footballs as baseballs in a sports store. There are 2.5 times as many baseballs as soccer balls in the same store.

(a) If the number of soccer balls is s, express the number of footballs and baseballs in terms of s.

(b) If there are 105 footballs in the store, how many soccer balls are there?

28. Alyssa bought 4 pens and 3 markers for $\$d$. If 1 pen cost $\$p$, find the cost of 1 marker.

29. Three times n is p. What is one-third of n in terms of p?

8 Algebraic Expressions

8.2 Simplifying Algebraic Expressions

Basics

1. Simplify the following expressions.

   (a) $2(m + m)$ _____

   (b) $2m \times m$ _____

   (c) $8e - 2e$ _____

2. Simplify the following expressions.

   (a) $8s + 3 - 2s$ _____

   (b) $y + 28 + y$ _____

   (c) $8g - 2 - 5g$ _____

   (d) $5h + 4 - 3h - 1$ _____

   (e) $3k + 8 - 3 - k$ _____

   (f) $10 - 5t - 2 - 2t$ _____

   (g) $4c - c^2 + 32 - c + 18$ _____

3. Simplify the following expressions.

   (a) $\frac{9}{3}p$ _____

   (b) $\frac{3}{5}k \times \frac{2}{3}$ _____

   (c) $4g \times 5$ _____

Practice

4. Determine if the following expressions are equivalent or not.

   (a) $24a + 7$ and $7 + 24a$

   (b) $24a - 7$ and $7 - 24a$

   (c) $5r + 4 + r - 2$ and $1 + 7r + 1 - r$

   (d) $7 + k + 2 + 3k$ and $3k + 5 + 4 + 2k$

5. Use the distributive property to rewrite these expressions as a sum or difference.

   (a) $6 \times (3b - 4)$ _____

   (b) $7(2e + 3)$ _____

   (c) $2(\frac{1}{3} + j)$ _____

   (d) $15(3 - p)$ _____

6. Use the distributive property to rewrite these expressions as products.

   (a) $14y + 7$ _____

   (b) $9x + 12$ _____

   (c) $15 - 10l$ _____

   (d) $16m - 4$ _____

   (e) $32w + 8 - 4w - 4$ _____

> **Challenge**

7. Simplify.

   (a) $8(a + b) - 2(3a + 2b)$

   (b) $3(c + d) - c^2 \times c + 2c$

8. Yara is 8 years old. Yara's mother is $4n$ times older than Yara. How old was Yara's mother 5 years ago in terms of n?

9. A chair costs $m and a table costs $100 more than the chair. Mila buys 2 tables and 8 chairs.

   (a) How much do the chairs and tables cost Mila? Express the amount in terms of m.

   (b) Mila has $1,000. If each chair costs $25, how much does she have left after paying for the chairs and tables?

10. $\frac{1}{2}$ lb of mushrooms cost $n. How much did Colton pay for 2 oz of mushrooms in terms of n?

9 Equations and Inequalities

9.1A Algebraic Equations

Basics

1. Determine whether $r = 4$ is a solution to each equation.

    (a) $r + \frac{3}{4} = \frac{19}{4}$

    (b) $49 = 45 + r$

    (c) $88 = 92 - r$

    (d) $16 - r = \frac{12}{3}$

2. Determine whether $c = 9$ is a solution to each equation.

   (a) $3c = 27$

   (b) $4 = \frac{3}{4}c$

   (c) $63 = 7c$

   (d) $11.7 = 1.3c$

3. Determine whether $m = \frac{2}{5}$ is a solution to each equation.

   (a) $m + m = \frac{4}{10}$

   (b) $15m = 6$

(c) $40 = 20m$

(d) $\frac{4}{15} = \frac{2}{3}m$

(e) $20m - 4 = 10m$

4. Determine whether $d = 0.32$ is a solution to each equation.

(a) $1 - 2d = 0.64$

(b) $d^2 = 0.1024$

(c) $1 - d^2 \times 2 = 0.4754 + d$

9 Equations and Inequalities

9.1B Balancing Equations

Basics

5. Solve each equation and check your answer.

    (a) $q + 9 = 27$

    (b) $p + 3.2 = 9$

    (c) $y + \frac{3}{15} = \frac{4}{5}$

    (d) $t + 16.8 = 55$

6. Solve each equation and check your answer.

   (a) $p - 7 = 21$

   (b) $z - 93 = 128$

   (c) $y - \frac{3}{4} = \frac{3}{8}$

   (d) $s - 0.56 = 2.93$

   (e) $40 - v = 12$

7. Solve each equation and check your answer.

   (a) $100 = n + 37$

   (b) $3 = y + \frac{6}{5}$

   (c) $\frac{5}{8} = n - 3$

   (d) $1.84 = p + 0.98$

8. Solve each equation and check your answer.

   (a) $6n = 84$

   (b) $3n = 8.4$

(c) $4.2k = 25.2$

(d) $1.5s = 45$

9. Solve each equation and check your answer.

(a) $\frac{x}{5} = 7$

(b) $\frac{e}{7} = 12$

(c) $\frac{m}{3} = 1.02$

(d) $\frac{x}{4} = 11.2$

Practice

10. Solve each equation and check your answer.

(a) $\frac{2}{3}m = 12$

(b) $\frac{7}{8}e = 4.2$

(c) $\frac{3}{4}r = 75$

(d) $\frac{2}{5}d = \frac{1}{2}$

(e) $\frac{7}{9}b + 3 = 66$

11. Isaac received 29 baseball cards for his birthday to add to his collection. He now has 413 baseball cards. How many baseball cards did Isaac have before his birthday? Write an equation and solve.

12. One morning, a teacher corrected some math tests. In the afternoon, the teacher corrected 17 more math tests. At that point, there were 43 math tests corrected. How many math tests did the teacher correct in the morning?

13. Ellen had some fabric. After she used $2\frac{2}{3}$ yards to make a dress for her daughter, Ellen still had $3\frac{5}{8}$ yards left. How many yards of fabric did Ellen have before she made the dress? Write an equation and solve.

14. Melvin used $4\frac{2}{3}$ cups of milk in a soup recipe. He had $2\frac{3}{4}$ cups left. How much milk did Melvin have before he made the soup? Write an equation and solve.

15. Kai is $\frac{2}{3}$ as old as Rea. If Kai is 12 years old, how old is Rea? Write an equation and solve.

16. Margaret made 4 times as much money babysitting as her brother made. If Margaret earned $78 babysitting, how much money did her brother make?

17. Scientists were comparing the distance two frogs could jump. The first frog jumped $\frac{3}{4}$ as far as the second frog. If the first frog jumped 3 feet 6 inches, how far did the second frog jump?

18. A large bamboo plant is 3 times as tall as a small bamboo plant. If the large plant is 16.5 feet tall, how tall is the small bamboo plant?

19. Mr. and Mrs. Fujimoto drove from New York City to Toronto, which is about 348 miles. The distance that Mrs. Fujimoto drove on this trip was about $\frac{1}{3}$ as far as Mr. Fujimoto drove on this trip. About how many miles did Mr. Fujimoto drive?

20. Ms. Madani is four times as old as her son, Arman. Arman is 5 years older than his sister, Paula. Let y represent Paula's age, in years.

    (a) Express Ms. Madani's age in terms of y.

    (b) If Ms. Madani is 32 years old, how old is Paula?

21. A club knits caps for a charity. They knitted 26 more caps in August than they did in September. If the club knitted a total of 250 caps in August and September, how many caps did they knit in September?

22. ABC is an isosceles triangle. Write an equation to find the value of x and solve.

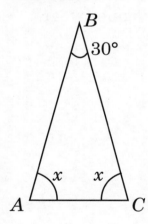

> **Challenge**

23. If $\frac{1}{4}$ of a number n is h, what is two times n?

24. Ani has a coin collection. In her collection, she has 8 more dimes than nickels. The value of all the dimes and nickels is $2.00. How many nickels does Ani have in her collection?

Equations and Inequalities

9.2A Algebraic Inequalities

Basics

1. Determine if 4 is a solution for the following inequalities.

   (a) $x > -3$

   (b) $x < 4\frac{1}{8}$

   (c) $x < 4$

2. Determine if -3 is a solution for the following inequalities.

   (a) $y > -5.06$

   (b) $y < -3$

   (c) $y > 3.73$

3. Determine if $-1\frac{1}{5}$ is a solution for the following inequalities.

   (a) $x > 1$

   (b) $x < -2$

   (c) $x > -1$

Equations and Inequalities

9.2B Graphing Inequalities Using a Number Line

Basics

4. Graph the solution for each of the inequalities.

(a) $x > -2$

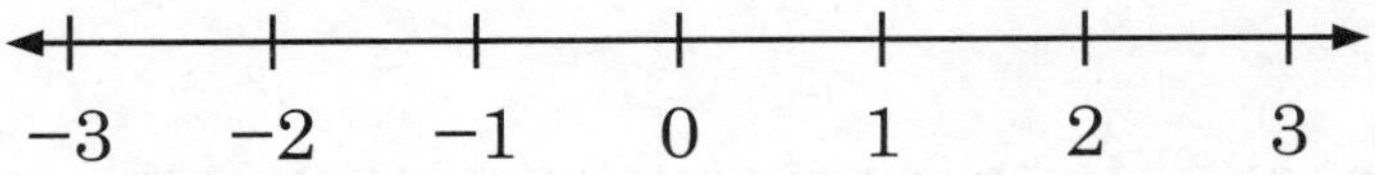

(b) $x \geq -2$

(c) $x < -4$

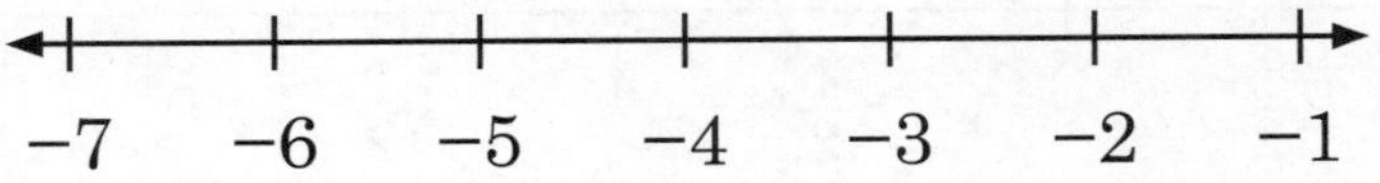

(d) $x \leq 8$

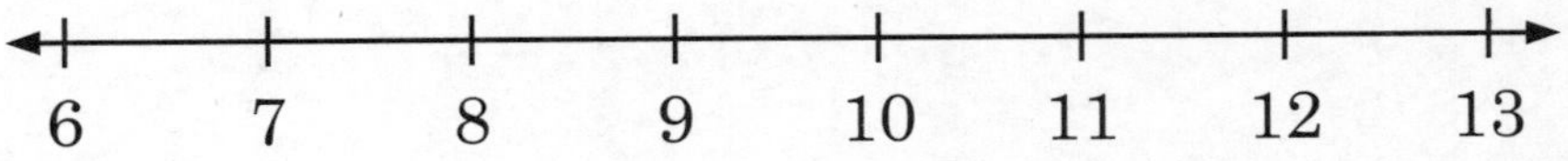

(e) $x \geq -2\frac{1}{2}$

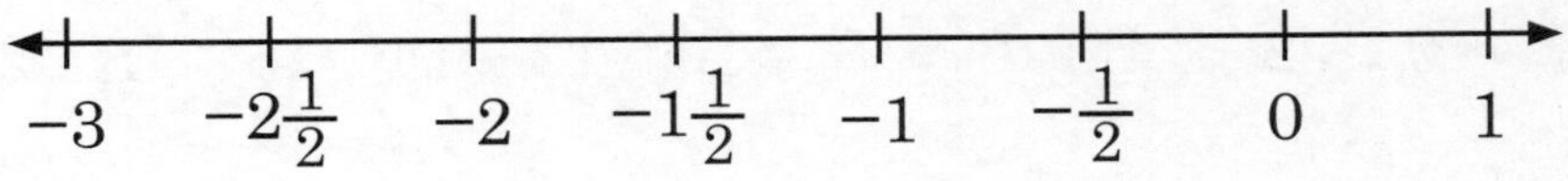

(f) $x < 3.5$

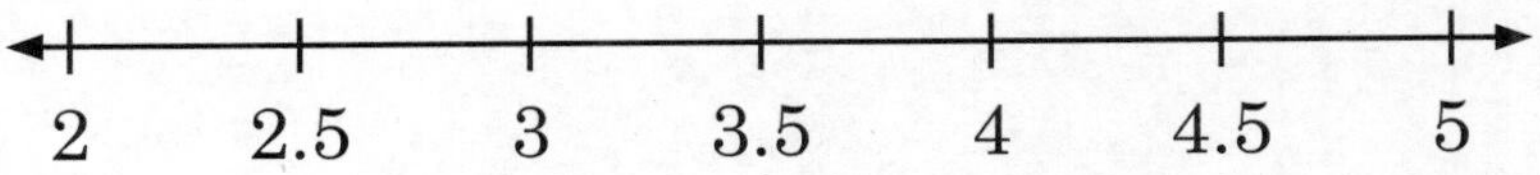

5. The Johnson family needs at least $800 to take a weekend trip.

   (a) Write an inequality to show this situation.

   (b) Graph the solution for the inequality on a number line.

   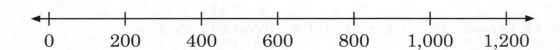

6. Write an inequality to represent each situation, and graph each solution on a number line.

   (a) Daniel is at least 4 feet tall.

   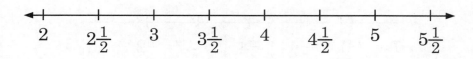

   (b) Samuel has some money. He has at most $18.

   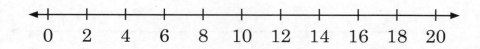

(c) Harry has more than 6.5 pounds of grapes.

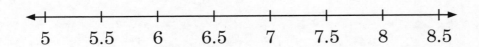

(d) The minimum speed on a highway is 45 miles per hour. The maximum speed is 65 miles per hour.

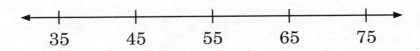

Practice

7. Each student in a sixth grade class spent at least two hours volunteering in the community.

(a) Write three different possible values for the number of hours spent volunteering by a student in the sixth grade class.

(b) Write an inequality to represent the number of hours spent volunteering by each student in the class.

(c) Graph the inequality on a number line.

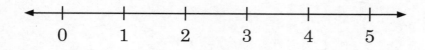

Challenge

8. For each full hour Linda spends walking dogs, she earns $16. She needs at least $128 to buy art supplies. How many hours does Linda need to work to pay for the art supplies? Write an inequality and solve.

9. In order to be allowed to ride a merry-go-round, a child must be at least 3 feet tall, but no taller than 4.5 feet. Write an inequality and draw a graph to represent this situation.

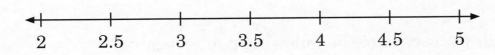

10 Coordinates and Graphs

10.1 The Coordinate Plane

Basics

1. Identify the *x*-coordinate for each point.

   (a) (3, −5) (b) (−2, 8)

   (c) (7, 0) (d) (0, 2.5)

2. Identify the *y*-coordinate for each point.

   (a) (0, −1.75) (b) (−12, 0)

   (c) (−3, 8) (d) $(4, -\frac{1}{2})$

3. In which quadrant is each of the following points located?

   (a) (−3.2, 4) (b) (1, 1)

   (c) $(-1\frac{1}{2}, -3\frac{3}{4})$ (d) (4, −3.6)

   (e) $(-3\frac{1}{3}, -2\frac{2}{3})$ (f) $(\frac{3}{4}, -1\frac{1}{2})$

Practice

4. Name the coordinates of points V, W, X, Y, and Z.

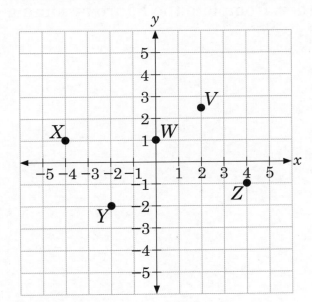

V _____

W _____

X _____

Y _____

Z _____

5. Number the x- and y-axes from 5 to −5. Plot and label each of the following points on the coordinate plane.

   (a) $R(-1.5, 3)$

   (b) $S(-2, -2)$

   (c) $T(1.5, 2)$

   (d) $U(0, 3)$

   (e) $W(1.5, -1)$

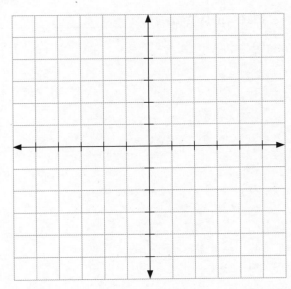

6. Plot the following points on the coordinate plane.

   P(3, 4), Q(–2, 4), R(–2, –2), S(3, –2)

   Connect the points in the given order. Complete the figure by joining the endpoints. Identify the polygon.

   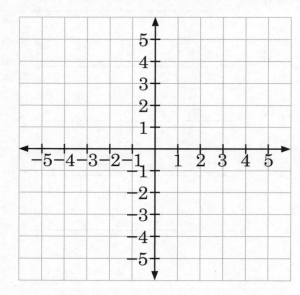

   The polygon is a _____.

Challenge

7. Susma is graphing right triangle DEF in a coordinate plane. Two of the vertices of the triangle are $D(4,-1)$ and $E(4,-3)$. The x-coordinate of point F is -2. What are the possible y-coordinates of point F?

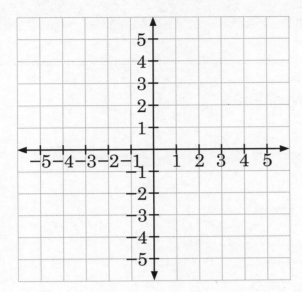

10 Coordinates and Graphs

10.2 Distance between Coordinate Pairs

Basics

1. Find the value of the following expressions.

    (a) $|15|$ _____

    (b) $|-8| - |-2|$ _____

    (c) $|5| + |-1|$ _____

    (d) $|-6.2| + |5|$ _____

2. Determine whether each pair of points can be used to form a vertical or horizontal line segment and if it is located in the same quadrant. If both points are in the same quadrant, tell which quadrant.

    (a) $A(4, 8)$ and $B(4, -3)$

    (b) $C(-2, 3)$ and $D(-4, 3)$

    (c) $E(-1, 2)$ and $F(-1, -8)$

    (d) $G(1, -5)$ and $H(-5, -5)$

3. Plot the points and draw a line segment. Determine whether each pair of points is in the same quadrant and find the distance in units between the points.

	Located in the same quadrant?	Distance between points
(a) $P(-1, 3)$ and $Q(-8, 3)$		
(b) $R(-1, -5)$ and $S(-6, -5)$		
(c) $T(3, -2)$ and $U(6, -2)$		
(d) $V(-1\frac{1}{2}, 6)$ and $W(1\frac{1}{2}, 6)$		

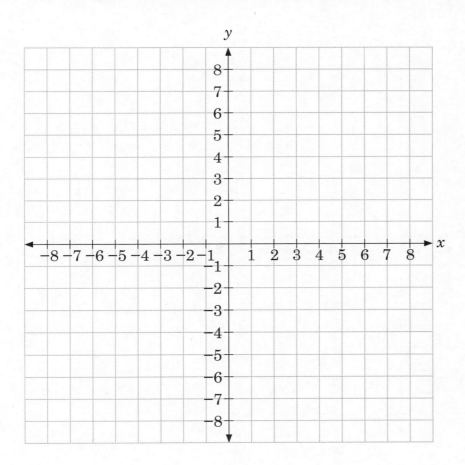

Coordinates and Graphs — 10.2 Distance between Coordinate Pairs

4. Plot the points and draw a line segment. Determine whether each pair of points is in the same quadrant and find the distance in units between the points.

	Located in the same quadrant?	Distance between points
(a) $H(2, -1)$ and $I(2, -4)$		
(b) $J(-3, 4)$ and $K(-3, -2)$		
(c) $L(5, 7)$ and $M(5, 4)$		
(d) $N(-4, -1)$ and $O(-4, -5)$		

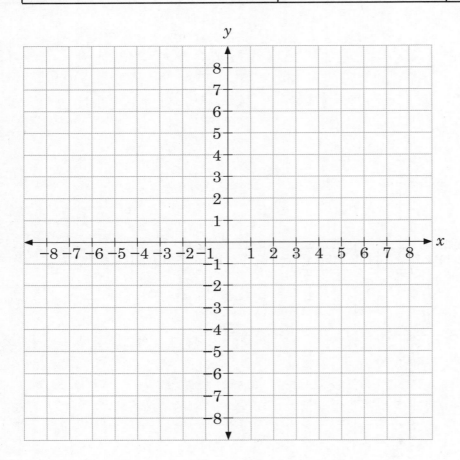

5. What is the distance between point W(7, 3) and point Z(−4, 3)? Take one unit to be 1 m.

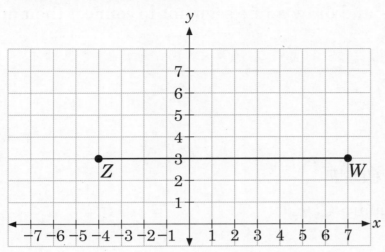

6. Draw points C(−6, 2) and D(−6, −6). What is the distance between the points? Take one unit to be 1 cm.

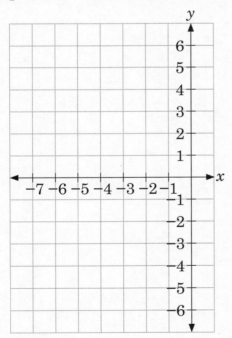

Practice

7. Plot the following points and draw a line segment to connect them in the order given.

   C (−7, 6)
   D (−2, 6)
   E (−2, 1)
   F (−7, 1)

   Find the area of square CDEF.

   Find the perimeter of square CDEF.

   Each unit is 1 cm.

   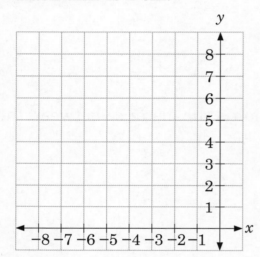

8. Draw rectangle *FGHI* with the following coordinates. Take one unit to be 1 cm.

   $F(-3, -4)$, $G(-3, 2)$, $H(2, 2)$, $I(2, -4)$

   (a)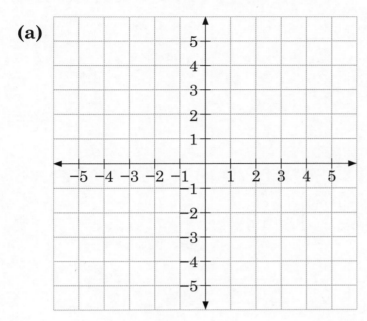

   (b) Find the perimeter of rectangle *FGHI*. Write equations for finding the length using the distance from the axes and the absolute value symbols.

   (c) Find the area of rectangle *FGHI*.

9. The points *W*, *X*, and *Y* are the vertices of a square. Plot a point *Z* in the diagram to complete the square *W*, *X*, *Y*, and *Z*. What are the coordinates of *Z*?

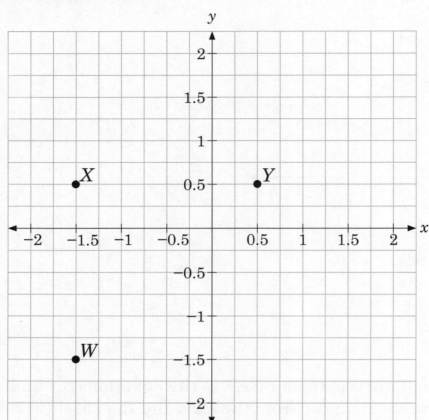

10. For the pair of points shown in each diagram below, determine if the given pairs of points are the endpoints of a horizontal line segment, vertical line segment, or diagonal line segment. If it is a horizontal or vertical line segment, determine the length. Take one unit in the coordinate plane to be 1 m.

(a)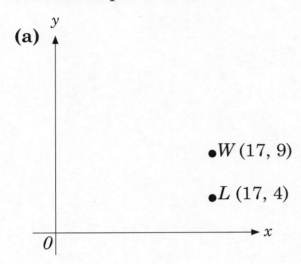

(b)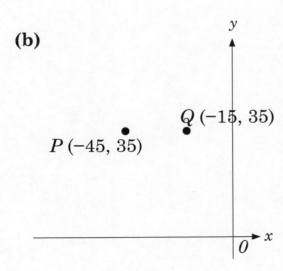

(c)

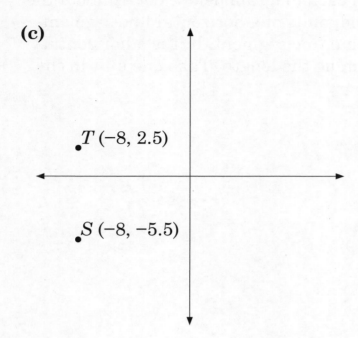

(d)

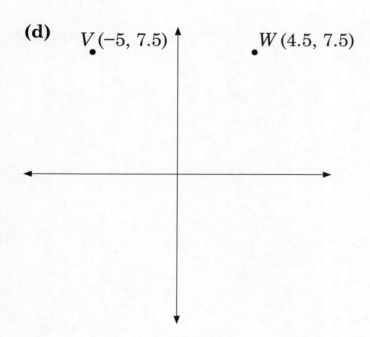

11. Determine whether a line segment drawn between each pair of points is vertical or horizontal, and if each pair of points is located in the same quadrant.

	Vertical or Horizontal Line Segment	Located in the same quadrant?
(a) $P(4, 0)$ and $Q(-9, 0)$		
(b) $R(3, 2\frac{1}{2})$ and $S(3, 6\frac{1}{4})$		
(c) $T(-3.5, 8)$ and $U(3, 8)$		
(d) $V(-5, -7.8)$ and $W(5, -7.8)$		

12. Find the distance between each pair of points. Take one unit on the coordinate plane to be 1 cm.

	Distance Between the Points
(a) $A(-7.5, 6)$ and $B(-5, 6)$	
(b) $C(-3\frac{3}{4}, 8)$ and $D(-1, 8)$	
(c) $E(110, 27)$ and $F(99, 27)$	
(d) $G(-55, -75)$ and $H(-55, -125)$	
(e) $I(-11\frac{5}{8}, 85\frac{1}{2})$ and $J(-11\frac{5}{8}, 89)$	
(f) $K(99, -45\frac{5}{8})$ and $L(99, -15\frac{1}{8})$	

13. **(a)** Find the distance between point $Y\left(-3\frac{1}{2}, 2\frac{1}{4}\right)$ and point $Z\left(-3\frac{1}{2}, -7\frac{3}{4}\right)$.

    Take one unit on the coordinate plane to be 1 cm.

    (b) Use the points Y and Z to form a line segment for one side of a square. Give the coordinates of two other vertices of the square. Is there more than one possible answer?

    (c) Find the area and perimeter of the square you constructed.

14. Find the area and perimeter, in units, of a rectangle with these coordinates.

    $K(-30, 15)$, $L(-30, -40)$, $M(10, -40)$, $N(10, 15)$

Challenge

15. Find the vertices of a rectangle located by following the clues.

  (a) The area of the rectangle is 48 units^2.

  (b) Half of the area is below the x-axis.

  (c) Two-thirds of the area is to the right of the y-axis.

  (d) The coordinates for one point are (4, 4).

  (e) All lines are either horizontal or vertical.

16. Find the coordinates of a square by following the clues. Is there more than one square that fits the criteria?

  (a) The coordinates are whole numbers only.

  (b) The four vertices of this square each lie in a different quadrant.

  (c) The length of one side is 8 units.

  (d) The area of this square located in the fourth quadrant is 20 units^2.

  (e) The area of this square located in the second quadrant is 12 units^2.

  (f) The lines are either horizontal or vertical.

17. Find the area of triangle *ABC* whose vertices have coordinates at point *A*(0, 4), point *B*(−4, −4), and point *C*(4, −4).

10 Coordinates and Graphs

10.3A Independent and Dependent Variables

Basics

1. Identify the independent and dependent variable in each scenario.

   (a) Jennifer picks b quarts of blueberries and makes p pints of jam.

   (b) Joyce scores p points at a school basketball game after making b baskets.

   (c) Jessica earns s stickers for c chores she completes.

2. In each of the following scenarios, identify the dependent and independent variable. Show your answers in a table.

   (a) At a picnic, the more hamburgers that are eaten, h, the more hamburgers buns, b, are used.

   (b) A runner burns 11.25 calories, c, for each minute, m, she runs.

   (c) An interior decorator is planning to paint a living room. For every gallon of paint, g, she can paint w square feet of wall.

   (d) A copy machine can make c copies in every h hours.

10 Coordinates and Graphs

10.3B Representing Relationships between Variables

Basics

3. Amy was born 52 years after her grandfather. Let A be Amy's age in years, and G be her grandfather's age in years.

   (a) Make a table to show Amy's age when her grandfather's age is 55 through 60.

   (b) Write an equation to show the relationship between the ages of Amy and her grandfather.

   (c) How old will Amy be when her grandfather is 75?

4. Carter makes $15.25 per hour at his part-time job. Let t be the total amount of money he earns in dollars. Let h be the number of hours Carter works.

   (a) Which variable is independent, and which is dependent?

   (b) Write an equation to show the relationship between the number of hours Carter works and the total amount of money he earns in dollars.

   (c) Create a table to show the total amount of money Carter earns if he works 8, 9, 10, 11, and 12 hours.

   (d) Carter saves $\frac{3}{5}$ of his earnings. About how many hours will Carter have to work in order to save $300?

5. Mrs. Ahmed took a cab home from work. The cost was $2.60 plus an additional $0.46 per mile.

   (a) Write an equation to represent the total amount, t, in dollars after a specific number of miles, m.

   (b) Identify the independent and dependent variables.

   (c) Make a table to show the total cost for 10 to 15 miles.

   (d) Mrs. Ahmed lives 16 miles from work. How much will this cab ride cost?

10 Coordinates and Graphs

10.3C Observing Relations between Variables with Graphs

Basics

6. The following table shows the relationship between time, d days, and distance, m miles, driven by Jenna on a trip. Write an equation to model the relationship between the time and distance.

   (a)

Number of Days d	Miles Driven m
1	350
2	700
3	1,050
4	1,400

   (b) Represent this relationship on the graph below.

   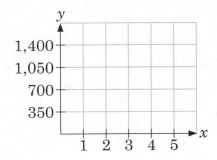

Coordinates and Graphs — 10.3 Changes in Quantities

7. Albert drives at a constant rate of 60 miles an hour. Let h be the number of hours Albert drives. Let m be the number of miles he drives.

   (a) Identify the independent and dependent variable in this scenario.

   (b) Make a table to show how many miles Albert drives between 3 and 10 hours.

   (c) Write an equation to show the relationship between m and h.

   (d) On the graph below, show the relationship between m and h.

   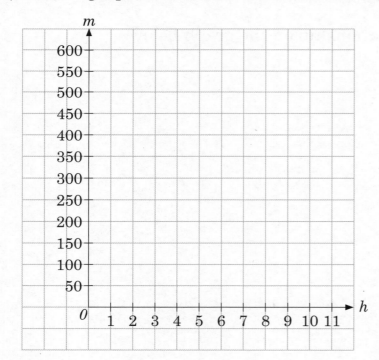

8. Canvas comes on bolts of 39 yards. Let x be the number of bolts and y be the total number of yards.

   (a) Write an equation that relates x and y.

   (b) Create a table to show the relationship between x and y. Show at least 5 values.

   (c) Represent the relationship on the graph below.

   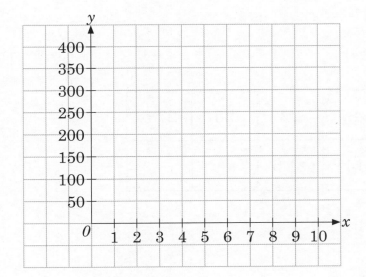

9. Complete each table. Write an equation to show the relationship between the variables in each table. Assume that y is the dependent variable in each case.

(a)

x	1	2	3	4	5
y	3.5	7	10.5		

(b)

x	3	5	9	15	30
y	10	12	16		

(c)

x	11	14.5	22	25	27.5
y	10.5	14	21.5		

10. Write an equation that represents the relationship between the two variables.

(a)

x	y
2	6
5	15
13	39
17	51

(b)

s	t
4	8
5.5	11
9	18
22.2	44.4

(c)

p	q
5	250
12	600
17	850
23	1,150

Practice

11. The internet service at the airport costs $11 to sign on, and an additional $2.50 per half hour. Let h represent the amount of time Frances used the internet and t represent the total cost in dollars. Write an equation that represents this scenario. Create a table that shows 0 to 3.5 hours, including half hours, of internet usage at the airport, and a graph that shows the relationship between these variables.

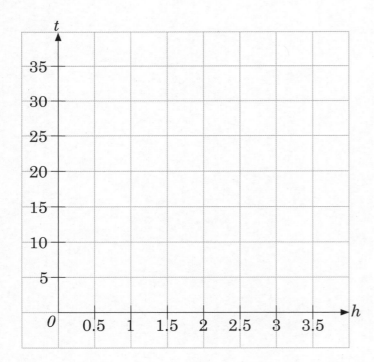

12. The Big City Bicycle Messenger Service charges an $8 fee, and an additional $3.25 per mile to make a delivery. Let m represent the miles biked and c represent the total cost of a delivery in dollars. Write an equation, complete the table, and create a graph that shows the relationship between the variables.

Number of Miles Biked m	1	3			14	18	20
Total Cost in Dollars c			30.75	47.00			

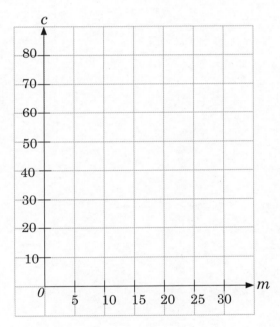

13. A country club charges a yearly membership fee of $250. It costs an additional $45 to play 1 round of golf. Let g represent the number of rounds of golf and let t represent the total amount of money, in dollars. Write an equation that describes the relationship between the total cost and the rounds of golf. Create a table and graph to show the amount per year that one would pay to play between 1 and 25 rounds of golf.

14. Olga is building a patio. The cost will be $22 per square foot. The length of the patio must be 12 ft. Let f represent the number of square feet of the patio and t represent the total cost, in dollars, of building the patio.

(a) Create a table that shows how much it will cost Olga to build a patio that is between 100 and 200 square feet. Choose 5 values for f.

(b) Write an equation to represent the relationship.

(c) Using the values you found in (a), graph the relationship between the number of square feet and the total cost of the patio.

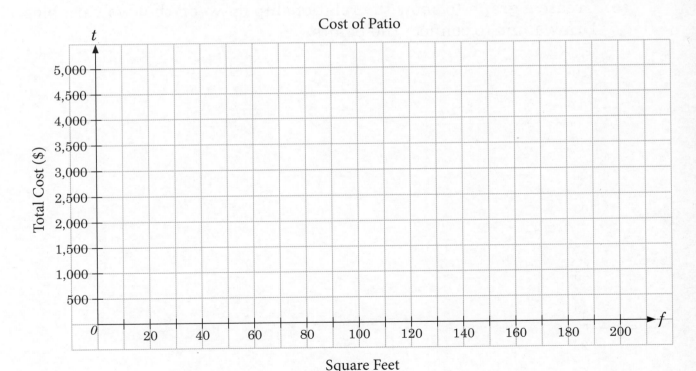

15. A hose can fill a pool at a rate of 8 gallons every half hour. The table below shows the number of gallons, g, that will go into the pool every hour, h. Complete the table.

(a)

Number of Hours h	Number of Gallons g
0.5	8
1	
2	
4	
7.5	
10.5	

(b) Write an equation that shows the relationship between the number of hours the hose is running and the number of gallons in the pool.

(c) Create a graph to show the relationship between the two variables. Draw a line to connect the points.

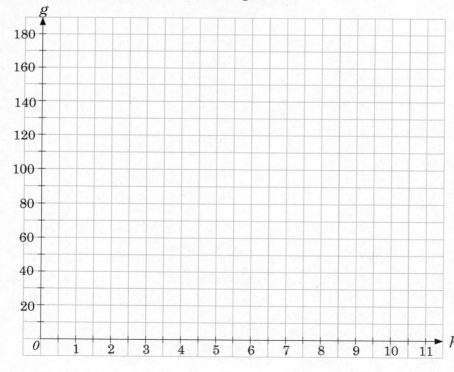

Challenge

16.

x	2	3	4	5	6	8		12
y	15	20	25		35		55	

(a) Graph the relationship between x and y for those points where you have both values.

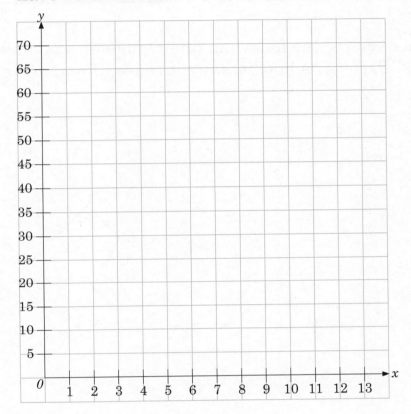

(b) Use the graph to find the missing values and complete the table.

(c) Write an equation that represents the relationship.

(d) Use the equation to find x when $y = 100$.

11 Area of Plane Figures

11.1 Area of Rectangles and Parallelograms

Basics

1. Determine whether each of the following line segments is a height of parallelogram *ABCD*. If it is a height, state its base. If not, explain why.

   (a) *MN*

   (b) *CQ*

   (c) *AD*

   (d) *OP*

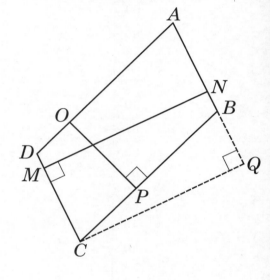

2. Name a line segment which gives the height corresponding to base *LM* in parallelogram *JKLM*. Then name another base and corresponding height for the parallelogram.

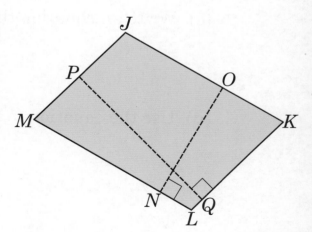

72　Area of Plane Figures — 11.1 Area of Rectangles and Parallelograms

3. In each parallelogram, identify a base and the corresponding height, and then determine the area.

(a)

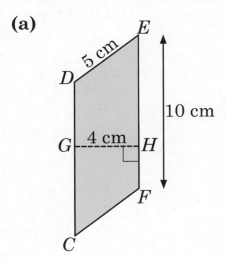

(b)

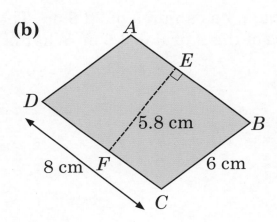

(c)

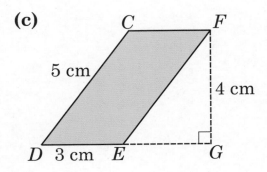

Practice

4. The figure below is made up of two parallelograms. Find its area.

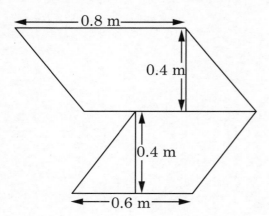

5. A mural, in the shape of a parallelogram, has an area of 22.8 m². The distance between the two longer sides of the mural is 3.8 m. What is the length of each longer side?

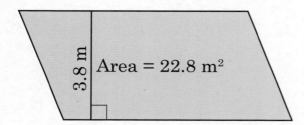

6. The length of one side of parallelogram *TUVW* is 9.6 cm and the area is 67.2 cm². Find the perpendicular distance between the given pair of opposite sides.

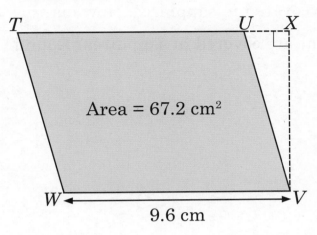

7. An advertising company created a billboard in the shape of a parallelogram. The ink colors for the parallelogram are red, blue, and yellow, and the areas in those colors, respectively, are in the ratio of 4 : 3 : 1. What area of the billboard is blue?

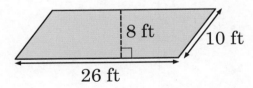

8. A jeweler created a parallelogram-shaped pendant. $\frac{1}{4}$ of the pendant is covered in emeralds, $\frac{1}{6}$ of it is covered in rubies, $\frac{1}{12}$ of it is covered in diamonds, and the rest of it is covered in sapphires. How many square centimeters of the pendant are covered in sapphires? Round to the nearest tenth.

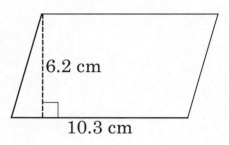

9. Draw Parallelogram *ABCD* on the coordinate graph below. Then find the area of the parallelogram.

    Point *A* is located at (−5, −2)
    Point *B* is located at (2, −2)
    Point *C* is located at (4, 4)
    Point *D* is located at (−3, 4)

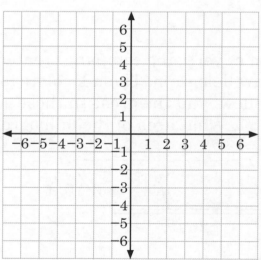

Challenge

10. A tailor is cutting a $1\frac{1}{4}$ yd by $\frac{2}{3}$ yd piece of rectangular cloth into two pieces by cutting on the diagonal. Find the area of each of the pieces.

11. Alyssa has entered a contest to design a logo for an advertising class she is taking. She drew two parallelograms, adjacent to each other. The smaller parallelogram, shown below, has an area of 13.5 cm². The ratio of the base of the smaller parallelogram to that of the larger parallelogram is 1 : 2. The ratio of the height of the smaller parallelogram to that of the larger parallelogram is 2 : 3. What is the total area of the two parallelograms?

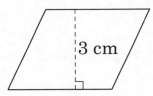

11 Area of Plane Figures

11.2A Finding Area of a Triangle

Basics

1. Determine whether each of the following line segments is a height of triangle *LMN*. If it is a height, state its corresponding base. If not, explain why.

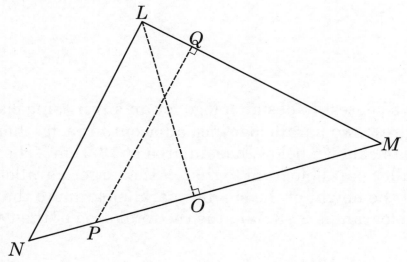

(a) *OL*

(b) *PQ*

(c) *MN*

2. Label a base and a corresponding height for each of the following triangles.

(a)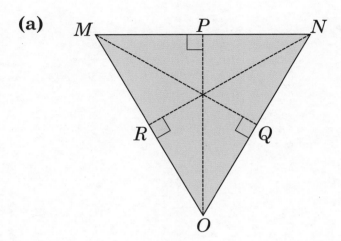

(b)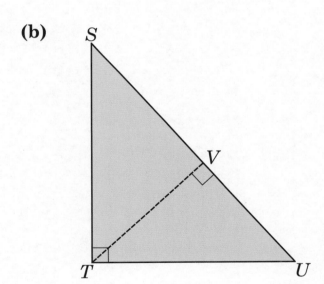

3. Identify a base and a corresponding height of triangle *DEF*, and then find its area.

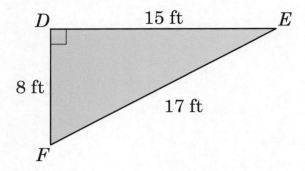

4. Find the area of each triangle.

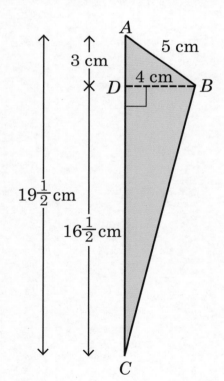

Practice

5. Find the area of the shaded triangles. Round your answer one decimal place.

(a)

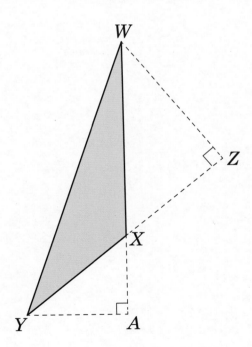

(b) *AY* is 2 in, *XZ* is 3.2 in, *YW* is 6.8 in, *WZ* is 3 in, and *ZY* is 6.1 in.

Area of Plane Figures — 11.2 Area of Triangles

6. A park has a triangular garden with a long southern edge and a northern vertex, as shown. The area of this garden is 140 ft². If the distance from the vertex V to the southern edge is 8 ft, what is the length of this edge?

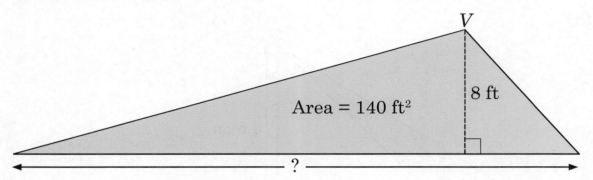

Challenge

7. In the figure to the right, all lines are straight lines. *HJ* and *GI* intersect at right angles at point *K*. *GK* = 2 in, *HK* = $2\frac{3}{4}$ in, *IK* = $1\frac{3}{4}$ in, and *JK* = $3\frac{3}{8}$ in.

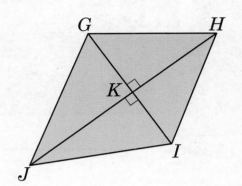

What is the difference between the area of triangle *GHK* and the area of triangle *HIK*?

11 Area of Plane Figures

11.2B Areas Involving Parallelograms and Triangles

Basics

8. The area of parallelogram $FGHI$ is 38.25 cm². What is the area of triangle FGH, rounded to two decimal places?

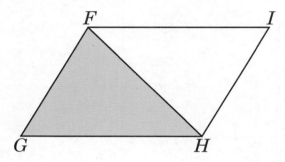

9. The picture below shows a banner designed by a sixth grade class. The banner will be cut from a piece of rectangular fabric which measures 3 ft × 4 ft. How much fabric will be left over?

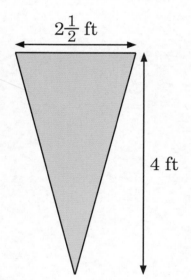

10. Find the area of the shaded triangle.

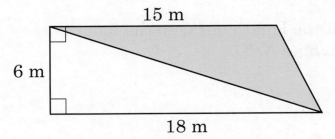

11. The area of parallelogram EFGH is 96 cm². What is the length of x?

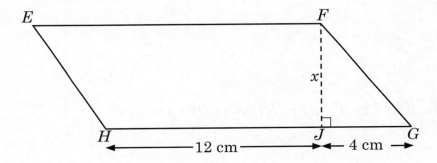

Practice

12. MNOP is a square. The ratio of the length of PQ to the length of OQ is 3 : 1. Find the area of triangle NPQ.

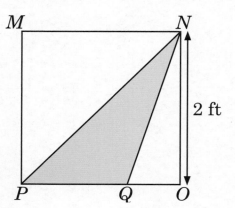

13. The area of triangle KLM is 25 cm². What is the area of parallelogram MNOP?

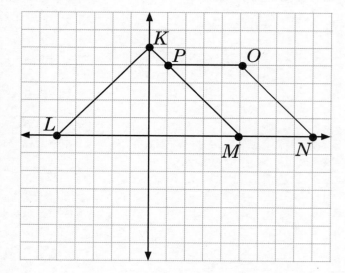

Challenge

14. The flag below contains several smaller shapes. Find the combined area of figures *A* and *B*.

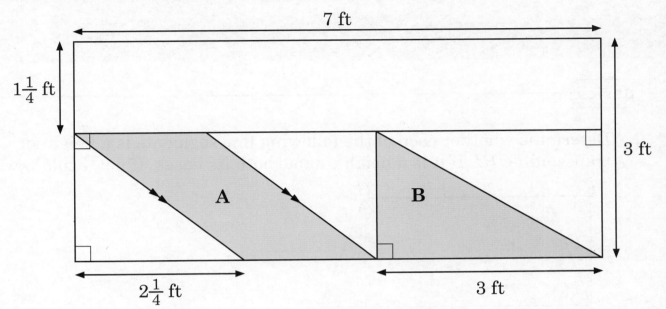

11 Area of Plane Figures
11.3 Area of Trapezoids

Basics

1. Determine whether each of the following line segments is a height of trapezoid *FGHI*. If it is a height, name both its bases. If not, explain why.

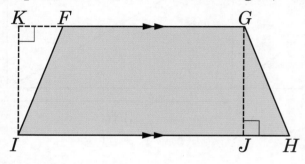

(a) *FG*

(b) *GH*

(c) *GJ*

(d) *IK*

2. Identify the bases and the height of each trapezoid, and then find its area.

(a)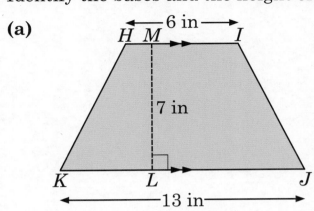

(b)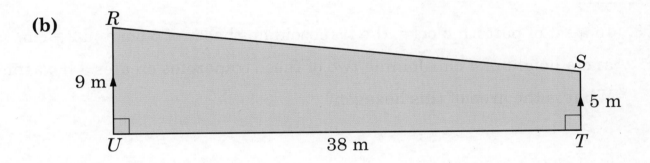

(c)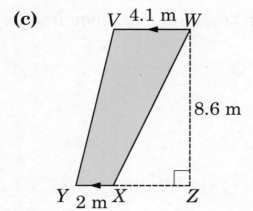

Practice

3. In a set of pattern blocks, the trapezoid has bases of 5 cm and $2\frac{1}{2}$ cm, and a height of 4 cm. Joining two of these trapezoids creates a hexagon. What is the area of this hexagon?

4. Trapezoid *ABCD* has an area of 20 square cm. On a coordinate grid, the coordinates of three of the vertices are *A* (1,1), *B* (3,1), and *C* (5, −3). Use the grid below to draw the trapezoid. Show your work for how you determined the coordinates of D.

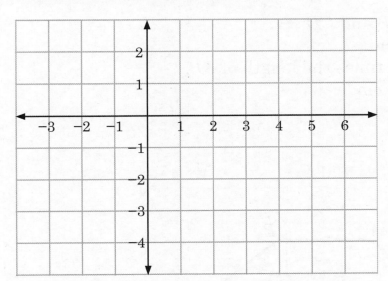

Challenge

5. Alyssa drew this trapezoid. Each square in the grid below is 0.5 cm by 0.5 cm.

The parallel sides are *AB* and *CD*.
The length of *AB* is 2.5 cm.
The length of *CD* is four times the length of *AB*.
The height of *ABCD* is 3 cm.

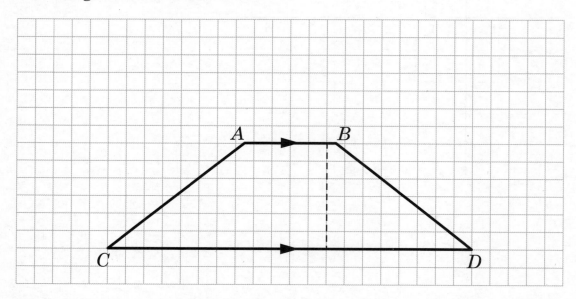

Draw a different trapezoid that has the same area. Each square in the grid below is 0.5 cm by 0.5 cm.

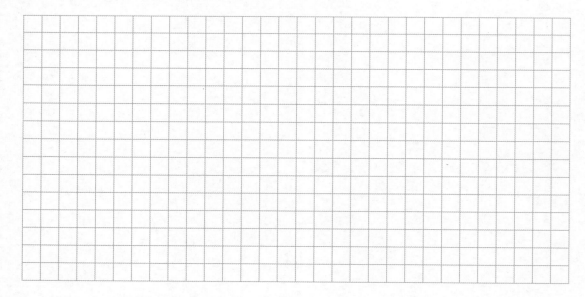

12 Volume and Surface Area of Solids

12.1A Cubes and Cuboids

Basics

1. Find the volume of this cube. Express the answer in simplest form.

   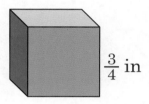

2. Find the volume of the rectangular prism. Express the answer as a mixed number in simplest form.

   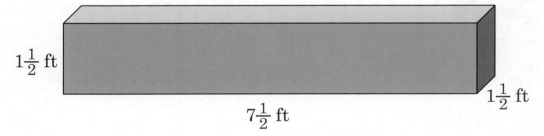

3. Find the volume of the rectangular prisms.

   (a)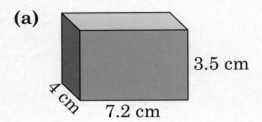

   (b)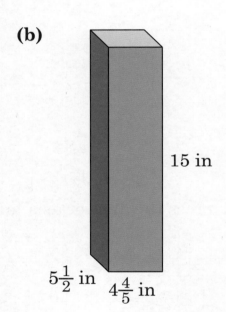

4. Find the volume of a rectangular prism with a base area of 42 ft², and a height of $16\frac{2}{3}$ ft.

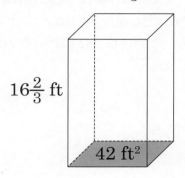

5. Find the width of the cuboid.
   Volume = 410.4 cm³

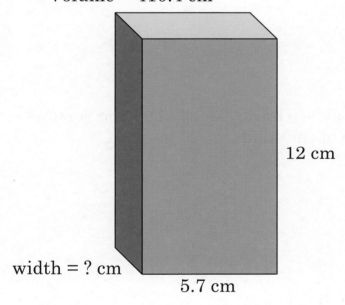

Practice

6. Two rectangular prisms have the same volume of 72 cm³. Rectangular prism A has a base of 9 cm² and is 4 times the height of rectangular prism B. All dimensions are whole numbers of centimeters. Give two possible sets of dimensions of rectangular prism B.

7. The solid below is made up of cuboids. The angles at all of the vertices are right angles. Find the volume of the solid.

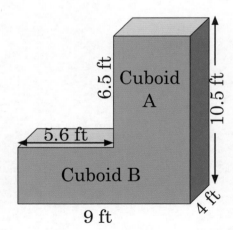

8. A paint vat presently holds 7.75 L of acrylic paint. It could hold 3.25 L more paint. If the vat is 25 cm long and 22 cm wide, what is the height of the vat?

9. Find the volume of the solid. The vertices are all right angles.

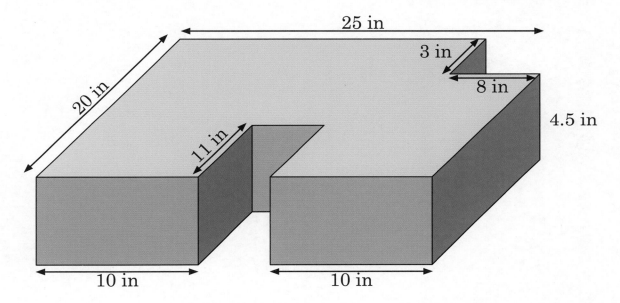

10. The length and height of each section of this solid are the same. Find the volume of this solid.

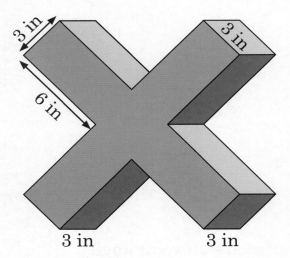

Challenge

11. Ms. Choi designs fancy aquariums for professional offices. Her most popular aquarium is 8 feet long, 2 feet wide, and 3 feet tall.

(a) What is the volume of Ms. Choi's most popular aquarium?

(b) The offices Ms. Choi designs aquariums for have different spaces allotted for their aquariums. Find three different sets of dimensions for aquariums which have the same volume as Ms. Choi's most popular aquarium. Each dimension must be a whole number of feet.

12. A wooden block is 12 cm long, 8 cm wide, and 5 cm high. A hole, with cross section of a square, is drilled through the entire length to form a hollow rectangular prism. The side length of the hole is 2.5 cm. What is the volume of the drilled block?

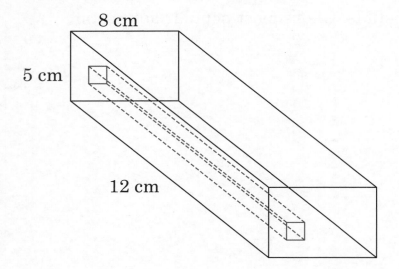

Volume and Surface Area of Solids

12.1B Volume of Liquids

Basics

13. A tank measuring 24 cm by 12 cm by 12 cm is $\frac{7}{8}$ filled with water. What is the volume of the water in the tank?

14. A cubical container with 4 cm edges is filled to the brim with water. If the water is poured from this container into an empty rectangular container measuring 12 cm by $2\frac{2}{3}$ cm by 4 cm, what fraction of the rectangular container will be filled?

15. When a tank is $\frac{1}{2}$ full it contains 45 liters of water. The area of the base is 450 cm². What is the height of the tank?

16. A rectangular container with a length of 10.5 cm, a width of 13 cm, and a height of 15.5 cm is filled with 1,500 cm³ of water. What is the volume of space that is not filled with water?

Practice

17. A tank measures 15 cm by 5 cm by 8 cm. After a period of evaporation, the container is $\frac{2}{3}$ filled. If the remaining water is poured into a smaller container and fills it completely, what could be the dimensions of the smaller container if all dimensions are in whole numbers of cm?

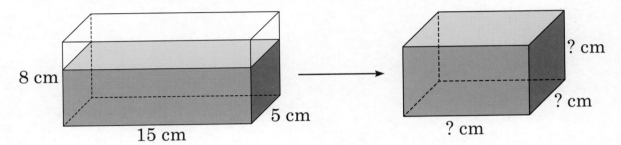

18. A rectangular tank measuring 60 cm long, 40 cm wide, and 50 cm high is $\frac{3}{4}$ full of water. Water is being let out at a rate of 10 liters per minute. How long will it take for the tank to empty?

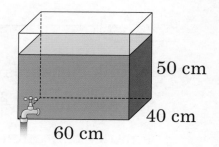

19. A 90 cm by 30 cm by 45 cm rectangular tank is being filled at a rate of 18 liters per minute. How long will it take to fill the tank to capacity?

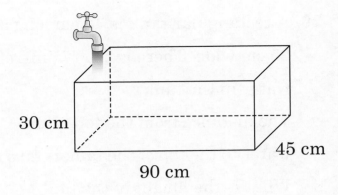

20. A rectangular tank is 42 cm long and 25 cm wide. There is 5 L 775 mL of water in the tank. The distance from the top of the water to the top of the tank is 19.5 cm. What is the height of the tank?

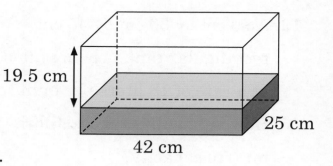

21. Sydney needs a fish tank that is at least 150 L, but takes up as little room on her desk as possible. Tank A measures 50 cm by 45 cm by 70 cm. Tank B measures 55 cm by 60 by 50 cm. Which fish tank should Sydney buy? Explain your answer.

Challenge

22. A swimming pool, 15 ft long, 12 ft wide, and 6 ft deep is $\frac{3}{5}$ filled with water. A hose is adding water to the pool at a rate of 1 ft³ per minute. The pool has a leak and is draining water at a rate of 0.25 ft³ per minute. Find the new water level in the pool after 30 minutes.

23. A rectangular fish tank is 16 cm long and 24 cm wide. It is filled with water to a height of 20 cm. A metal cube with a side length of 12 cm is put into the fish tank. What is the difference between the original height of the water and the new height of the water?

12 Volume and Surface Area of Solids

12.2A Surface Area of Rectangular Prisms

Basics

1. A cube has a surface area of 54 ft². What is the volume of the cube?

2. Find the surface area of each cuboid.

   (a)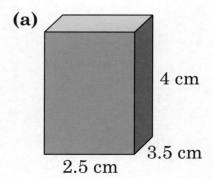

   4 cm, 3.5 cm, 2.5 cm

(b)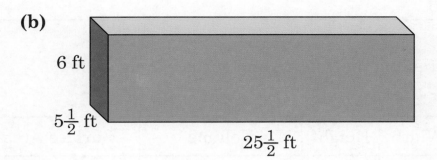

3. Find the surface area of the cube.

3.5 cm

Practice

4. Fill in the table by finding the volume and surface area of the rectangular prisms.

Prism	Length	Width	Height	Volume	Surface Area
a	2 m	1.5 m	4 m		
b	4.5 ft	3.5 ft	5 ft		
c	55 cm	35 cm	60 cm		
d	1.5 in	2 in	1.75 in		

5. Find the volume and surface area of the following cubes.

(a)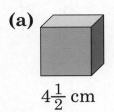
$4\frac{1}{2}$ cm

(b)
7 m

6. Ms. Castillo is shipping two packages. Box *A* measures 10 cm by 13 cm by 3.5 cm. Box *B* measures 10 cm by 8 cm by 5 cm. What is the difference between the surface area of the two boxes?

> **Challenge**

7. The following picture shows a solid consisting of two rectangular prisms on top of a larger rectangular prism.

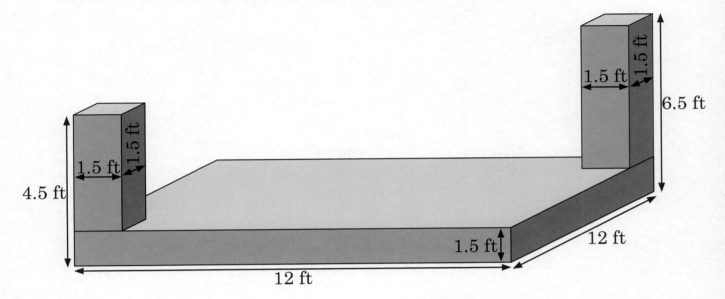

(a) Find the volume of the solid.

(b) Find the total surface area.

8. Which are NOT nets of cubes?

(a)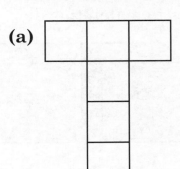

(b)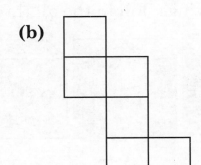

(c)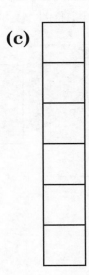

(d)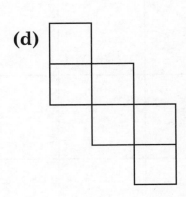

(e)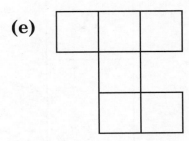

(f)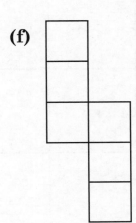

9. This is one die shown from different perspectives. The letters have been rotated in the drawings. Fill in the all three nets to match this die.

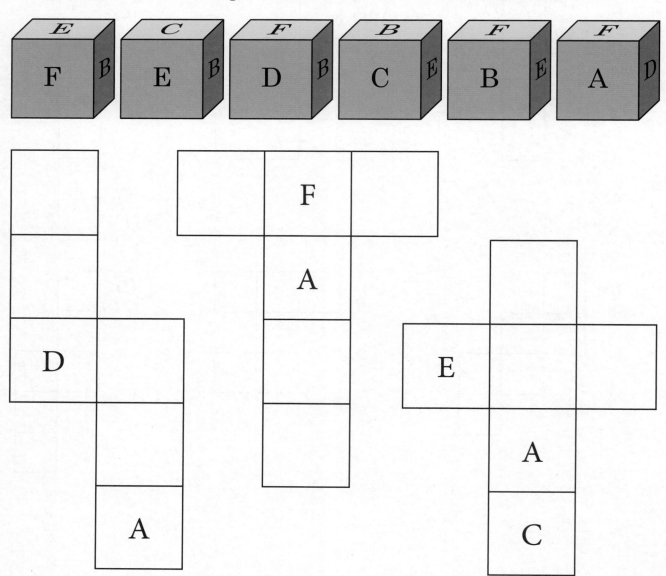

12 Volume and Surface Area of Solids

12.2B Surface Area of Triangular Prisms

Basics

10. A storage unit in the shape of a triangular prism is shown here. What is the surface area of the storage unit?

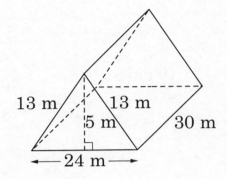

11. Find the surface area of this triangular prism.

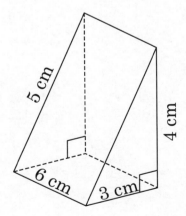

Practice

12. A chocolate shop uses triangular prism boxes. The boxes are crafted out of cardboard and then the rectangular faces are covered in fabric.

The fabric costs $0.02 per square cm. What is the cost for the fabric to produce one such box?

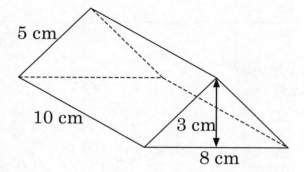

13. Consider the storage unit described in problem 10 on page 119. If the distance from the front base to the back base was $22\frac{1}{2}$ m rather than 30 m, what would be the surface area of the storage unit?

Volume and Surface Area of Solids
12.2C Metric Conversions and Volume

Basics

14. For the following cuboid, find:

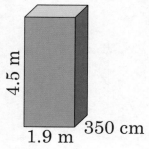

(a) the volume in both m^3 and cm^3.

(b) the surface area in both m^2 and cm^2.

Practice

15. Jasper's parents bought a rectangular sandbox that is 4 ft long, 3 ft wide, and 18 in tall for their backyard.

   (a) What is the maximum amount of sand, in ft³, the sandbox will hold?

   (b) Jasper's parents filled the sandbox with 75% natural sand and 25% colored sand. The sandbox is filled to three inches below the height of the box. How many cubic inches of each type of sand did they use?

16. Agnes's Hardware Store sells plastic tarps in different sizes. Order the tarps by area, from largest to smallest. Show your work.

Tarp	Area	Area in cm^2
Tarp A	120 cm^2	
Tarp B	1,350 cm^2	
Tarp C	2,345 mm^2	
Tarp D	13.45 m^2	

17. A company is painting three walls and the ceiling of a rectangular conference room. The room is 32 ft long, 28 ft wide, and 9 ft high. A window and large double doors are on one of the walls with the shorter length and this wall will not be painted. Each gallon of paint costs $18 and will cover 300 ft². If paint is only sold by the gallon, how much will the paint cost for this project?

13 Displaying and Comparing Data

13.1AB Statistical Questions & Measures of Center

Basics

1. For each of the following, determine whether or not the question is a statistical question. Give a reason for your answer.

    (a) How many total medals did the United States win in the 2016 Summer Olympics?

    (b) What are the favorite television shows of sixth grade students at your school?

    (c) What is the average price of video games at a certain store?

2. Below are the high temperatures in Phoenix, Arizona for one week in June, 2017. All temperatures are in degrees Fahrenheit.

    116, 112, 110, 108, 108, 109, 112

    Find the mean high temperature. Round to the nearest degree.

3. A basketball team's mean score for five games is 98. In the first four games, the team's scores were 104, 99, 95, and 88. What was the team's score in the fifth game?

4. The mean of six numbers is 184.2.
   The mean of five of the numbers is 178.
   What is the sixth number?

5. The mean of three numbers, d, e, and f, is 20.1.

   $e = 2d$. $f = d + e$.

   What is the difference between d and e?

6. The weight, in pounds, of nine packages are:

46, 29, 8, 78, 40, 37, 39, 66, and 80

(a) What is the mean weight of the nine packages?

(b) What is the median weight of the nine packages?

Practice

7. Antonio, Helen, Shivani, and Max are salespeople who get paid commissions based on their sales each week. A table illustrating the total sales for each of them for the month of April is shown below.

	Antonio	Helen	Shivani	Max
Week 1	$2,000	$1,800	$2,500	$1,200
Week 2	$3,200	$3,600	$2,800	$2,500
Week 3	$2,800	$3,000	$1,000	$1,800
Week 4	$5,000	$4,800	$3,000	$1,600

(a) What was the mean of all weekly sales in April?

(b) What was the median of all weekly sales that month?

(c) How would the mean sales average have changed if Max had not worked in April? Explain your reasoning.

8. Six two-digit numbers starting with 46 are arranged in increasing order. If the median of the six numbers is 58 and the mean is 63.5, what could the six numbers be? Show all of your work.

9. Find sets of integers that meet the requirements.

    (a) Three integers with a mean of 4 and a mode of 2

    (b) Three integers with a mean of 8 and a mode of 7

    (c) Three integers with a mean of 25 and the difference between the greatest integer and the least integer is 35

    (d) Four integers with a mean of 21, a median of 20, and the least integer is $\frac{1}{10}$ the greatest integer

10. Find the mean, median, and mode of the following set of numbers:

    $\frac{3}{4}, \frac{2}{3}, \frac{1}{4}, \frac{1}{3}, \frac{1}{8}, \frac{1}{2}$

> Challenge

11. To calculate the batting average of a baseball player, statisticians start by finding the ratio of the number of hits to the number of times at bat for that player. They then round that decimal number to the nearest thousandth. The tables below contain data needed to solve the following problems.

(a) What was Babe Ruth's mean batting average for the three-year period of 1926, 1927, and 1928?

Year	Times at Bat	Hits
1926	495	184
1927	540	192
1928	536	173

(b) What was Jose Altuve's mean batting average for the three-year period of 2015, 2016, and 2017?

Year	Times at Bat	Hits
2015	638	200
2016	640	216
2017	590	204

12. A sixth grade student was analyzing his quiz scores during a marking period. His scores were: 20, 85, 90, 82, 94, 88, 82, 86, 95.

  (a) What was the student's mean score on these quizzes, rounded to the nearest tenth?

  (b) What was the student's median score on these quizzes?

  (c) Is the mean or the median more representative of the students' average score? Explain your answer.

  (d) What would be the student's mean score, rounded to the nearest tenth, if the teacher were to allow the student to drop the lowest quiz score?

13 Displaying and Comparing Data
13.2A Dot Plots

Basics

1. The heights of basketball players trying out for a professional basketball team are shown below.

   6'6" (6 ft 6 in)   5'9"   6'9"   6'8"   6'10"   6'6"   6'3"   6'7"   6'11"
   6'4"   6'6"   6'8"   6'5"   5'10"   6'8"   6'7"   6'6"   6'5"   6'7"   6'9"

   (a) Find the mean height of the players trying out, in feet.

   (b) Display the data in a dot plot.

   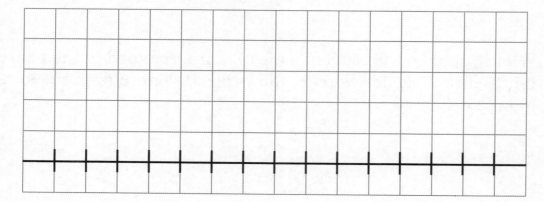

   (c) Describe briefly the distribution of the data.

2. XYZ Company is keeping track of the number of overtime hours worked by its employees. The chart below shows the data collected during a one-week period.

Number of overtime hours worked	0	1	2	3	4	5	6
Number of employees	2	4	5	3	3	2	1

(a) Find the mean number of overtime hours worked by XYZ Company employees during this week.

(b) Display the data in a dot plot.

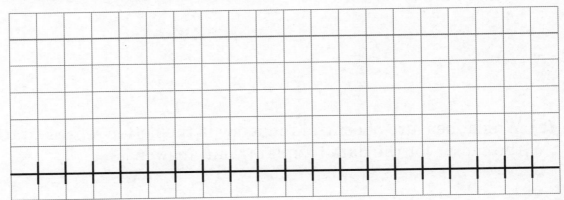

Practice

3. The dot plot below shows the typing speed in words per minute of a group of sixth grade students.

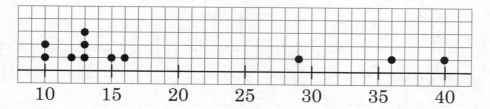

(a) How many students are in the group?

(b) What is the mean of the data? Round to one decimal place.

(c) What is the median of the data?

(d) What fraction of the students type fewer words per minute than the mean number of words?

(e) Which measure of central tendency is the better representation of the center for the data? Explain your answer.

13 Displaying and Comparing Data

13.2B Histograms

Basics

4. The following list shows the age, rounded to the nearest number of years, of people in the food court of a mall one morning.

0	54	1	1	11	0	62	23	2	12	8	3
1	10	9	1	46	4	1	9	5	6	5	8
2	3	48	2	67	27	6	33	4	7	7	2
53	3	66	5	2	6	3	35	64	19	17	59
26	24	31	42	3	29	31	65	28	2	35	3
14	15	18	18	19	20	13	20	22	31	32	34
31	25	27	34	41	32	29	12	11	12	14	25

(a) Explain why drawing a histogram to represent the data is better than drawing a dot plot.

(b) Create a grouped frequency of uniform intervals for the data.

(c) Draw a histogram in intervals of ten to represent the data.

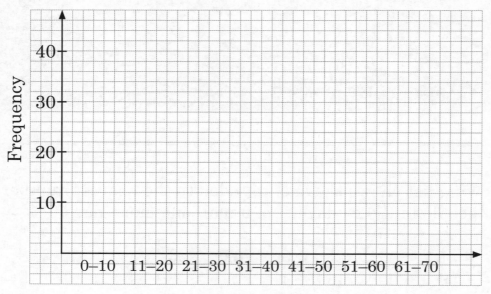

(d) What fraction of the people in the food court were between eleven and twenty years old?

(e) What percentage of the people in the food court were at least forty-one years old? Round your answer to a tenth of a percent.

> **Practice**

5. Mr. Ikeda, a math teacher, purchased a fitness tracker to log the number of steps he took every day for three weeks. The following table shows the number of steps (s) taken each day.

Day	Steps
Saturday	10,420
Sunday	10,109
Monday	8,675
Tuesday	7,854
Wednesday	6,043
Thursday	4,879
Friday	3,050
Saturday	10,056
Sunday	10,625
Monday	7,879
Tuesday	7,015
Wednesday	5,867
Thursday	5,475
Friday	3,128
Saturday	10,835
Sunday	10,521
Monday	8,561
Tuesday	7,298
Wednesday	5,106
Thursday	3,012
Friday	2,961

(a) Create a grouped frequency table of uniform intervals of 2,000 for the data.

(b) Create a histogram for the data.

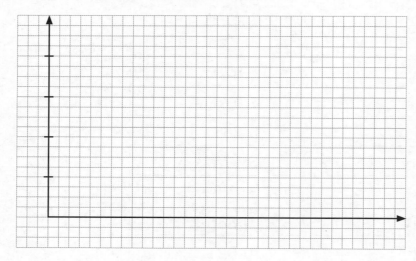

(c) Which interval best describes the center of the number of steps taken by Mr. Ikeda each day?

6. The histogram below shows the sales prices, in dollars, of video games sold in a game store in one hour on Black Friday.

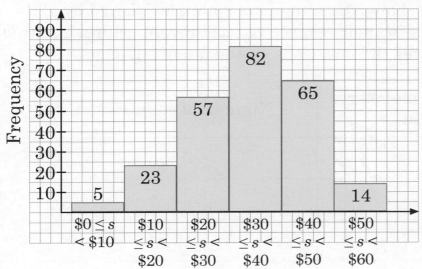

(a) What percent of the video games sold cost more than $30? Round your answer to a tenth of a percent.

(b) Describe the shape of the histogram.

Challenge

7. The following table shows the number of countries visited by 120 travel club members in their respective most recent cruise. Use this information to answer the questions on the following page.

Fraction of Club members (out of 120)	Number of countries visited
$\frac{1}{40}$	6
$\frac{1}{40}$	7
$\frac{1}{30}$	8
$\frac{1}{10}$	9
$\frac{7}{60}$	11
$\frac{1}{6}$	12
$\frac{7}{30}$	13
$\frac{3}{20}$	14
$\frac{1}{15}$	15
$\frac{1}{12}$	16

(a) Determine the number of club members according to the number of countries visited.

(b) Organize the data using a grouped frequency table.

(c) Draw a histogram to present the data.

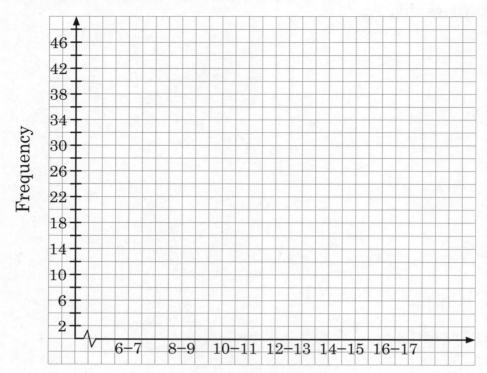

(d) Describe the shape of the histogram.

(e) The travel agent will use this data to plan future cruises. If he earns money on the number of passengers he books on a cruise, what would be the minimum number of countries he should plan the cruise to visit? Explain your answer.

13 Displaying and Comparing Data

13.3A Range

Basics

1. Kawai practiced his guitar for the following number of minutes each day during the first week of January: 7, 10, 25, 20, 35, 5, 30. During the second week of January his practice minutes each day were: 15, 18, 20, 15, 25, 26, 28.

   (a) Find the mean and median of each set of data.

   (b) Kawai's guitar teacher has told him that consistency in practice time is important. Would the guitar teacher be happier to see Kawai's practice time record for week 1 or week 2? Explain your answer.

2. The table below shows the amount of money saved by two sisters over an eight-week period.

	Week 1	Week 2	Week 3	Week 4	Week 5	Week 6	Week 7	Week 8
Holly	$3.25	$5.00	$8.80	$2.05	$5.25	$9.75	$1.50	$7.55
Jasmine	$4.95	$5.25	$6.05	$4.75	$5.70	$5.25	$5.25	$5.85

(a) Which girl saved more? How much more?

(b) What was the range of Holly's saving?

(c) What was the range of Jasmine's saving?

13 Displaying and Comparing Data
13.3B Mean Absolute Deviation

Basics

3. Consider the following set of data: 1, 3, 4, 5, 5, 6.

   (a) Find the mean.

   (b) Draw a dot plot to represent the data distribution.

   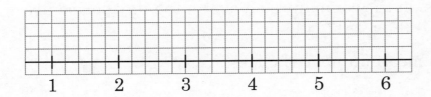

   (c) Draw a number line, labeling the mean, to help find the distance of data values from the mean.

   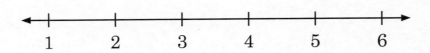

(d) Make a table to compute the absolute deviation in the data and find the sum of the absolute deviation.

(e) Find the mean absolute deviation (MAD) of this set of data. Round to two decimal places.

Practice

4. A small company is paying to run advertisements on Google. The number of clicks on the company's ad over a ten day period are 12, 21, 39, 13, 41, 36, 49, 37, 42, 45.

   (a) What is the mean number of clicks?

   (b) Make a table to compute the absolute deviations in the data. Then find the mean absolute deviation of the number of clicks. What does the MAD mean for this data set?

(c) Draw a dot plot to represent this data distribution.

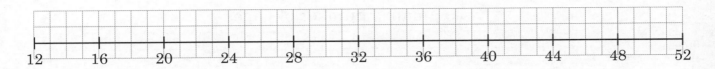

(d) Is the mean of this data a good indicator of a typical number of clicks? Explain your answer.

5. Imani and Logan kept track of the number of pages they read each day in their required reading books for seven days. The data from their reading logs is shown below.

Day Number	Imani's number of pages	Logan's number of pages
1	24	9
2	21	10
3	25	18
4	24	16
5	21	20
6	20	40
7	19	41

(a) What is the mean number of pages in Imani's and Logan's data?

(b) Draw dot plots of Imani's and Logan's number of pages.

Imani's number of pages

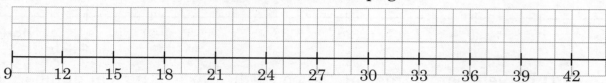

Logan's number of pages

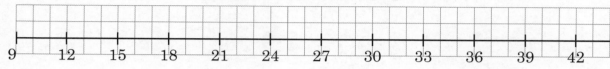

(c) Without doing any calculations, for which distribution do you think the mean would give a better indicator of a typical value?

(d) Calculate the MAD of Imani's and Logan's data. Round to two decimal places.

(e) Justify your answer in (c).

13 Displaying and Comparing Data

13.3C Interquartile Range

Basics

6. The average monthly rainfall, in inches, of three California cities is shown below.

	January	February	March	April	May	June
Sacramento	3.62	3.46	2.76	1.14	0.67	0.2
San Francisco	4.49	4.45	3.27	1.46	0.71	0.16
Bakersfield	1.14	1.22	1.22	0.51	0.2	0.08

	July	August	September	October	November	December
Sacramento	0.04	0.04	0.28	0.94	2.09	3.17
San Francisco	0.06	0.08	0.2	1.1	3.15	4.57
Bakersfield	0	0.04	0.08	0.31	0.63	1.02

For each city, find:

(a) the mean average monthly rainfall. Round to two decimal places.

(b) the median. Round to two decimal places.

(c) the lower quartile and the upper quartile. Round to two decimal places.

(d) the interquartile range of the data. Round to two decimal places.

7. Find the five-number summary for each city in the prior problem.

13 Displaying and Comparing Data
13.3D Box Plot

Basics

8. The data below shows the number of minutes spent playing video games on a weeknight by a group of sixth grade students.

Number of Minutes							
15	10	22	0	10	25	60	45
30	30	27	0	10	40	30	20

(a) Find the five-number summary for the data set.

(b) Draw a box plot for the data. Then use the five-number summary and the IQR to describe the average number of minutes playing video games by these students.

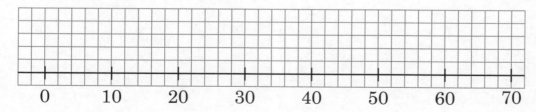

Challenge

9. Create two sets of data. Each set must have 7 data points. The mean of each set must be 10. Pick data points for the two sets of data such that their box and whisker plots are very different.

Answer Key

8.1 Writing and Evaluating Algebraic Expressions

1. (a) $s - 5$ (b) $15\frac{1}{2} - k$
   (c) $h + 2.25$ (d) $100 + m$

2. (a) $\frac{5c}{6}$ (b) $8.2e$
   (c) $1{,}000r$ (d) mc^2
   (e) $\frac{t}{12}$

3. $\frac{s}{10}$ and $\frac{1}{10}s$ are equivalent because $\frac{s}{10} = \frac{1 \times s}{10} = \frac{1}{10}s$.

4. (a) $\frac{5k}{3}, \frac{5}{3}k$ (b) $\frac{4c}{2}, \frac{4}{2}c$

5. (a) $3h - 5$
   (b) $\frac{a}{5} + 10$ or $\frac{1}{5}a + 10$
   (c) $\frac{3}{10}k + \frac{2}{5}$ or $\frac{3k}{10} + \frac{2}{5}$
   (d) $\frac{1}{2} - \frac{3}{5}t$ or $\frac{1}{2} - \frac{3t}{5}$
   (e) $7a + c$
   (f) $7n + 10p$

6. The difference is $(8 - k)$ minutes.

7. (a) $2p$ (b) $\frac{2}{3}p$
   (c) $p - 2$

8. $23 - p$ apps

9. $3b + 2p$ dollars

10. $\frac{c-2}{3}$ cookies

11. $(3s + 2c)$ grams

12. w^3 cm

13. $24b + n$ basketball cards

14. (a) 13   (b) 15
    (c) 81   (d) 2

15. (a) 6   (b) 12
    (c) $16\frac{2}{3}$ (d) 0.7

16. (a) 25   (b) 32
    (c) $1\frac{1}{2}$ (d) 15
    (e) 17

17. (a) 1   (b) 7
    (c) $\frac{5}{14}$ (d) $\frac{5}{14}$
    (e) 0

18. (a) $\frac{1}{8}$ (b) $\frac{3}{16}$
    (c) $9\frac{1}{16}$ (d) $\frac{11}{30}$

19. (a) 0.27   (b) 0.2025

20. (a) 8   (b) $1\frac{1}{2}$

21. (a) 12, 24, 36, 48, 60
    (b) $\frac{f}{12}$ flower arrangements
    (c) 15 flower arrangements

22. (a) $(b - 3)$ cm   (b) 29 cm

23. (a) $d + 75$ (b) 79 years old

24. (a) $g \times 8.6$ (b) $8.6g$ m^2
    (c) 67.08 m^2

25. (a) $2.5c$ dogs   (b) 10 dogs

26. (a) $w = \frac{7}{3}l$ or $2\frac{1}{3}l$
    (b) 12 m

27. (a) footballs: 7.5s
       baseballs: 2.5s
    (b) 14 soccer balls

28. $m = \frac{(d - 4p)}{3}$

29. $\frac{p}{9}$

8.2 Simplifying Algebraic Expressions

1. (a) $4m$ (b) $2m^2$
   (c) $6e$

2. (a) $6s + 3$ (b) $2y + 28$
   (c) $3g - 2$ (d) $2h + 3$
   (e) $2k + 5$ (f) $8 - 7t$
   (g) $3c - c^2 + 50$

3. (a) $3p$ (b) $\frac{2}{5}k$
   (c) $20g$

4. (a) equivalent (b) not equivalent
   (c) equivalent (d) not equivalent

5. (a) $18b - 24$ (b) $14e + 21$
   (c) $\frac{2}{3} + 2j$ (d) $45 - 15p$

6. (a) $7(2y + 1)$ (b) $3(3x + 4)$
   (c) $5(3 - 2l)$ (d) $4(4m - 1)$
   (e) $4(7w + 1)$

7. (a) $2a + 4b$ (b) $5c + 3d - c^3$

8. $(32n - 5)$ years old

9. (a) $\$(10m + \$200)$
   (b) $\$550$

10. $\$\frac{n}{4}$ or $\$\frac{1}{4}n$

9.1 Equations

1. (a) $r = 4$ is a solution
   (b) $r = 4$ is a solution
   (c) $r = 4$ is a solution
   (d) $r = 4$ is not a solution

2. (a) $c = 9$ is a solution
   (b) $c = 9$ is not a solution
   (c) $c = 9$ is a solution
   (d) $c = 9$ is a solution

3. (a) $m = \frac{2}{5}$ is not a solution
   (b) $m = \frac{2}{5}$ is a solution
   (c) $m = \frac{2}{5}$ is not a solution
   (d) $m = \frac{2}{5}$ is a solution
   (e) $m = \frac{2}{5}$ is a solution

4. (a) $d = 0.32$ is not a solution
   (b) $d = 0.32$ is a solution
   (c) $d = 0.32$ is not a solution

5. (a) $q = 18$ (b) $p = 5.8$
   (c) $y = \frac{9}{15}$ or $\frac{3}{5}$ (d) $t = 38.2$

6. (a) $p = 28$ (b) $z = 221$
   (c) $y = \frac{9}{8}$ or $1\frac{1}{8}$ (d) $s = 3.49$
   (e) $v = 28$

7. (a) $63 = n$ (b) $1\frac{4}{5} = y$
   (c) $3\frac{5}{8} = n$ (d) $p = 0.86$

8. (a) $n = 14$ (b) $n = 2.8$
   (c) $k = 6$ (d) $s = 30$

9. (a) $x = 35$ (b) $e = 84$
   (c) $m = 3.06$ (d) $x = 44.8$

10. (a) $m = 18$ (b) $e = 4.8$
    (c) $r = \frac{300}{3} = 100$
    (d) $d = 1\frac{1}{4}$ (e) $b = 81$

11. 384 baseball cards

12. 26 math tests

13. $6\frac{7}{24}$ yards of fabric

14. $7\frac{5}{12}$ cups of milk

15. 18 years old

16. $19.50

17. 56 inches or 4 feet 8 inches

18. 5.5 feet

19. about 261 miles

20. (a) $4y + 20$ (b) 3 years old

21. 112 caps

22. $x = 75°$

23. $8h$

24. 8 nickels

9.2 Inequalities

1. (a) 4 is a solution for $x > -3$
   (b) 4 is a solution for $x < 4\frac{1}{8}$
   (c) 4 is not a solution for $x < 4$

2. (a) -3 is a solution for $y > -5.06$
   (b) -3 is not a solution for $y < -3$
   (c) -3 is not a solution for $y > 3.73$

3. (a) $-1\frac{1}{5}$ is not a solution for $x > 1$
   (b) $-1\frac{1}{5}$ is not a solution for $x < -2$
   (c) $-1\frac{1}{5}$ is not a solution for $x > -1$

4. (a) [number line: open circle at −2, arrow right]
   (b) [number line: closed circle at −2, arrow right]
   (c) [number line: open circle at −4, arrow left]
   (d) [number line: closed circle at 8, arrow left]
   (e) [number line: closed circle at $-2\frac{1}{2}$, arrow right]
   (f) [number line: open circle at 3.5, arrow left]

5. (a) $x \geq \$800$
   (b) [number line: closed circle at $800, arrow right]

6. (a) $d \geq 4$
   [number line: closed circle at 4 ft, arrow right]
   (b) $s \leq 18$
   [number line: open circle at $0, closed circle at $18]
   (c) $h > 6.5$
   [number line: open circle at 6.5 lbs, arrow right]
   (d) $45 \leq s \leq 65$
   [number line: closed circles at 45 and 65]

7. (a) Answers will vary.
   (b) $2 \geq h$
   (c)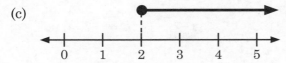

8. $h \geq 8$; at least 8 hours
9. $3 \leq t \leq 4.5$
   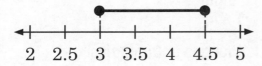

10.1 The Coordinate Plane

1. (a) 3   (b) -2
   (c) 7   (d) 0

2. (a) -1.75 (b) 0
   (c) 8   (d) $-\frac{1}{2}$

3. (a) 2nd Quadrant (b) 1st Quadrant
   (c) 3rd Quadrant (d) 4th Quadrant
   (e) 3rd Quadrant (f) 4th Quadrant

4. $V(2, 2.5)$, $W(0, 1)$, $X(-4, 1)$, $Y(-2, -2)$, $Z(4, -1)$

5.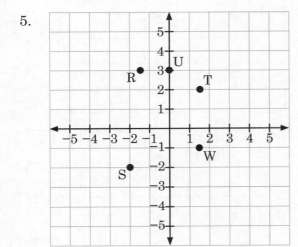

6. rectangle

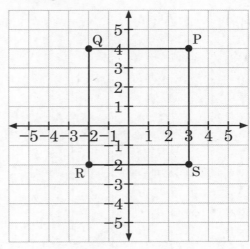

7. either -1 or -3

10.2 Distance between Coordinate Pairs

1. (a) 15   (b) 6
   (c) 6   (d) 11.2

2. (a) Vertical line segment, not in the same quadrant
   (b) Horizontal line segment, in the same quadrant, 2nd quadrant
   (c) Vertical line segment, not in the same quadrant
   (d) Horizontal line segment, not in the same quadrant

3. (a) yes, 7   (b) yes, 5
   (c) yes, 3   (d) no, 3

4. (a) yes, 3   (b) no, 6
   (c) yes, 3   (d) yes, 4

5. 11 m

6. 8 cm

159

©2017 Singapore Math Inc. Dimensions Math® Workbook 6B – Answer Key

7. Area = 25 cm²
   Perimeter = 20 cm

8. (a)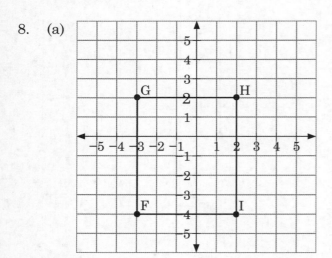

   (b) 22 cm   (c) 30 cm²

9. (0.5, −1.5)

10. (a) vertical; 5 m
    (b) horizontal; 30 m
    (c) vertical; 8 m
    (d) horizontal; 9.5 m

11. (a) horizontal; no
    (b) vertical; yes
    (c) horizontal; no
    (d) horizontal; no

12. (a) 2.5 cm   (b) $2\frac{3}{4}$ cm
    (c) 11 cm   (d) 50 cm
    (e) $3\frac{1}{2}$ cm   (f) $30\frac{1}{2}$ cm

13. (a) 10 cm
    (b) either $(6\frac{1}{2}, 2\frac{1}{4})$ and $(6\frac{1}{2}, -7\frac{3}{4})$
        or $(-13\frac{1}{2}, 2\frac{1}{4})$ and $(-13\frac{1}{2}, -7\frac{3}{4})$
    (c) Area: 100 cm²
        Perimeter: 40 cm

14. Area: 2,200 units²
    Perimeter: 190 units

15. (4, −4), (−2, −4), and (−2, 4)

16. (4, −5), (−4, −5), (−4, 3), and (4, 3), or
    (5, −4), (−3, −4), (−3, 4), and (5, 4)

17. 32 units²

10.3 Changes in Quantities

1. (a) Independent variable: number of quarts Jennifer picked, b
       Dependent variable: number of pints Jennifer can make, p
   (b) Independent variable: number of baskets Joyce made, b
       Dependent variable: number of points Joyce scored in the game, p
   (c) Independent variable: number of chores Jessica does, c
       Dependent variable: number of stickers Jessica earns, s

2.

	Independent	Dependent
(a)	h	b
(b)	m	c
(c)	g	w
(d)	h	c

3. (a)

Grandfather's age G	55	56	57	58	59	60
Amy's age A	3	4	5	6	7	8

   (b) $G = A + 52$ (c) 23 years old

4. (a) The number of hours Carter works, h, is the independent variable. The dependent variable is the total amount of money Carter earns, t.

   (b) $t = \$15.25h$

   (c)
No. of hours worked h	8	9	10	11	12
Total earned t	122	137.25	152.50	167.75	183

   (d) almost 33 hours

5. (a) $t = 2.60 + 0.46m$

   (b) The independent variable is the number of miles. The total cost of the cab ride is the dependent variable.

   (c)
Number of Miles m	Total Cost t, in dollars
10	7.20
11	7.66
12	8.12
13	8.58
14	9.04
15	9.50

   (d) $9.96

6. (a) $m = 350d$

   (b)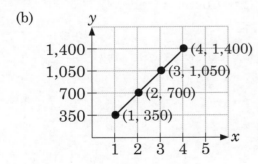

7. (a) The independent variable is the number of hours Albert drives. The dependent variable is the number of miles he drives.

   (b)
No. of Hours h	3	4	5	6	7	8	9	10
No. of Miles m	180	240	300	360	420	480	540	600

   (c) $m = 60h$

   (d)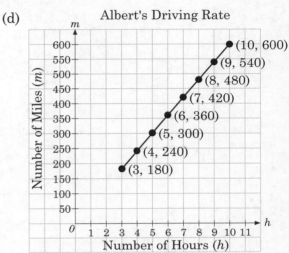

8. (a) $y = 39x$

   (b) Answers will vary.

   (c)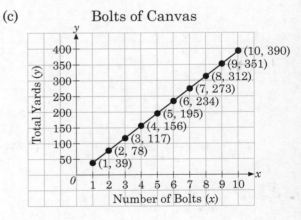

9. (a) table: 14, 17.5 $y = 3.5x$

   (b) table: 22, 37 $y = x + 7$

   (c) table: 24.5, 27 $y = x - 0.5$

10. (a) $y = 3x$

    (b) $t = 2s$

    (c) $q = 50p$

11. $t = 11 + 2.50h$

Time in Hours h	0	0.5	1	1.5	2	2.5	3	3.5
Total Cost in Dollars t	0	13.50	16.00	18.50	21.00	23.50	26.00	28.50

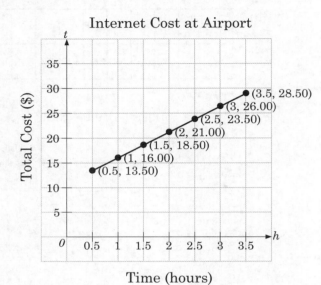

12. $c = 8 + 3.25m$

Number of Miles Biked m	Total Cost in Dollars c
1	11.25
3	17.75
7	30.75
12	47.00
14	53.50
18	66.50
20	73.00

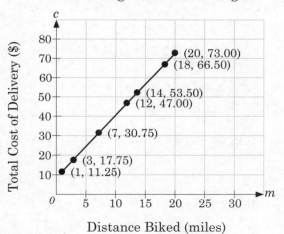

13. $t = 250 + 45g$

Rounds of Golf g	Cost in Dollars t	Rounds of Golf g	Cost in Dollars t
1	295	14	880
2	340	15	925
3	385	16	970
4	430	17	1,015
5	475	18	1,060
6	520	19	1,105
7	565	20	1,150
8	610	21	1,195
9	655	22	1,240
10	700	23	1,285
11	745	24	1,330
12	790	25	1,375
13	835		

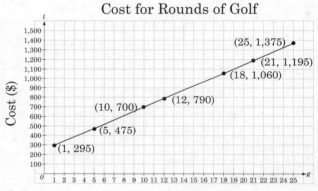

14. (a) Answers will vary.
    (b) $t = 22f$
    (c) Answers will vary.

15. (a)

Number of Hours h	Number of Gallons g
0.5	8
1	16
2	32
4	64
7.5	120
10.5	168

(b) $g = 16h$

(c) Rate a Hose Can Fill a Pool

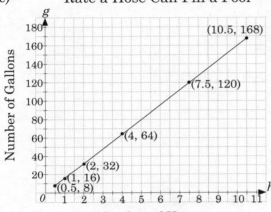

16. (a)

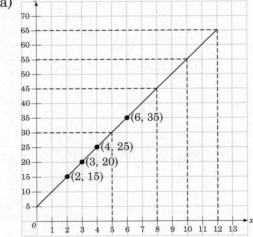

(b)

x	2	3	4	5	6	8	10	12
y	15	20	25	30	35	45	55	65

(c) $y = 5 + 5x$ or $y = 5x + 5$

(d) $x = 19$

11.1 Area of Rectangles and Parallelograms

1. (a) Yes, base CD and base AB.
   (b) Yes, base AB and base CD.
   (c) No. AD is a side of the parallelogram. It is neither perpendicular to BC nor CD.
   (d) Yes, base BC and base AD.

2. Height = NO
   Another base: KL, corresponding height PQ

3. (a) Base: EF
       Corresponding height: GH
       Area: 40 cm²
   (b) Base: AB
       Corresponding height: EF
       Area: 46.4 cm²
   (c) Base: DE
       Corresponding height: FG
       Area: 12 cm²

4. 0.56 m²

5. 6 m

6. 7 cm

7. 78 ft²

8. 31.9 cm²

9. 42 units²

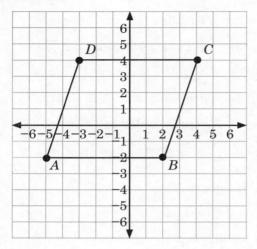

10. $\frac{5}{12}$ yd²

11. 54 cm²

11.2 Area of Triangles

1. (a) Yes, base: *MN*
   (b) No. Although *PQ* is perpendicular to base *LM*, it does not extend to the opposite vertex.
   (c) No. *MN* is not perpendicular to any base.

2. (a) Base *MN* — height *OP*
       Base *NO* — height *MQ*
       Base *MO* — height *NR*
   (b) Base *ST* — height *TU*
       Base *TU* — height *ST*
       Base *SU* — height *TV*

3. Base: *DF*, corresponding height: *DE*
   or
   Base: *DE*, corresponding height: *DF*
   Area: 60 ft²

4. *ABC*: 39 cm², *ABD*: 6 cm², *DBC*: 33 cm²

5. (a) 5.6 cm²
   (b) 4.4 in²

6. 35 ft

7. $\frac{11}{32}$ in²

8. 19.13 cm²

9. 7 ft²

10. 45 m²

11. 6 cm

12. $1\frac{1}{2}$ ft²

13. 16 cm²

14. $5\frac{11}{16}$ ft²

11.3 Area of Trapezoids

1. (a) No. *FG* is not a height. It is a base.
   (b) No. *GH* is not a height, as it is not perpendicular to the bases.
   (c) Yes. Bases: *FG* and *HI*
   (d) Yes. Bases: *FG* and *HI*

2. (a) 66.5 in²
       Bases *HI* and *JK*, Height *LM*
   (b) 266 m²
       Bases *RU* and *ST*, Height *TU*
   (c) 26.23 m²
       Bases *VW* and *XY*, Height *WZ*

3. 30 cm²

4.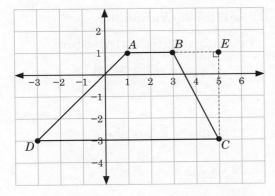

5. Answers will vary. A correct answer will show a trapezoid with an area of 18.75 cm².

12.1 Volume of Rectangular Prisms

1. $\frac{27}{64}$ in³

2. $16\frac{7}{8}$ ft³

3. (a) 100.8 cm³   (b) 396 in³

4. 700 ft³

5. 6 cm

6. Answers will vary.

7. 232.4 ft³

8. 20 cm

9. 1,894.5 in³

10. 243 in³

11. (a) 48 ft³
    (b) Answers will vary.

12. 405 cm³

13. 3,024 cm³

14. $\frac{1}{2}$ will be filled

15. 200 cm

16. 615.75 cm³

17. Answers will vary.

18. 9 minutes

19. 6.75 minutes

20. 25 cm

21. Fish tank A

22. 3.725 ft

23. 4.5 cm

12.2 Surface Area of Prisms

1. 27 ft³

2. (a) 65.5 cm²   (b) $652\frac{1}{2}$ ft²

3. 73.5 cm²

4. a. Volume = 12 m³
      Surface Area = 34 m²
   b. Volume = 78.75 ft³
      Surface Area = 111.5 ft²
   c. Volume = 115,500 cm³
      Surface Area = 14,650 cm²
   d. Volume = 5.25 in³
      Surface Area = 18.25 in²

5. (a) $91\frac{1}{8}$ cm³, $121\frac{1}{2}$ cm²
   (b) 343 m³, 294 m²

6. 81 cm²

7. (a) 234 ft³   (b) 408 ft²

8. (c) and (e) are not nets.

9.

	E							
	F		E	F	D		B	
	D	B		A		E	F	D
		C		C			A	
		A		B			C	

10. 1,620 m²

11. 84 cm²

12. $3.60

13. 1,245 m²

14. (a) 29.925 m³, 29,925,000 cm³
    (b) 61.9 m², 619,000 cm²

15. (a) 18 ft³
    (b) 6,480 in³ colored sand
        19,440 in³ natural sand

16. Tarp D, Tarp B, Tarp A, Tarp C

17. $108

13.1 Statistical Variability

1. (a) not a statistical question
   (b) statistical question
   (c) statistical question

2. 111 degrees Fahrenheit

3. 104 points

4. 215.2

5. 10.05

6. (a) 47 lbs
   (b) 40 lbs

7. (a) $2,662.50
   (b) $2,650
   (c) Answers will vary.

8. Answers will vary.

9. (a) 2, 2, 8   (b) 7, 7, 10
   (c) Answers will vary.
   (d) Answers will vary.

10. Mean: $\frac{7}{16}$
    Median: $\frac{5}{12}$
    Mode: There is no mode.

11. (a) 0.349
    (b) 0.332

12. (a) 80.2   (b) 86
    (c) Median score
    (d) 87.8

13.2 Displaying Numerical Data

1. (a) 6.5 ft
   (b)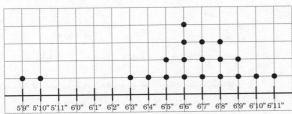
   (c) Answers will vary.

2. (a) 2.55 hours
   (b)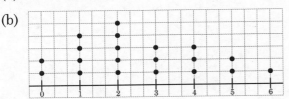

3. (a) 11 students
   (b) 18.8 words
   (c) 13
   (d) $\frac{8}{11}$
   (e) Median

4. (a) Answers will vary.
   (b) 0 – 10; 34
       11 – 20; 16
       21 – 30; 11
       31 – 40; 11
       41 – 50; 4
       51 – 60; 3
       61 – 70; 5
   (c)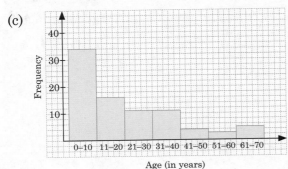

   (d) $\frac{4}{21}$
   (e) 14.3%

5. (a) $0 \leq s \leq 2{,}000$; 0
       $2{,}000 \leq s \leq 4{,}000$; 4
       $4{,}000 \leq s \leq 6{,}000$; 4
       $6{,}000 \leq s \leq 8{,}000$; 5
       $8{,}000 \leq s \leq 10{,}000$; 2
       $10{,}000 \leq s \leq 12{,}000$; 6
   (b)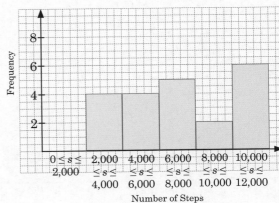
   (c) 6,000 to 8,000 steps

6. (a) 65.4%
   (b) Answers will vary.

7. (a) 3 members visited 6 countries
       3 members visited 7 countries
       4 members visited 8 countries
       12 members visited 9 countries
       14 members visited 11 countries
       20 members visited 12 countries
       28 members visited 13 countries
       18 members visited 14 countries
       8 members visited 15 countries
       10 members visited 16 countries

(b) 6 – 7; 6
8 – 9; 16
10 – 11; 14
12 – 13; 48
14 – 15; 26
16 – 17; 10

(c)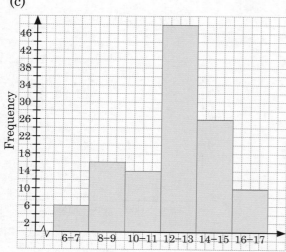

(d) Answers will vary.

(e) Answers will vary.

13.3 Measures of Variability and Box Plots

1. (a) Week 1: Mean $18\frac{6}{7}$, Median 20

    Week 2: Mean 21, Median 20

   (b) Week 2

2. (a) Holly, $0.10
   (b) $8.25   (c) $1.30

3. (a) Mean: 4
   (b)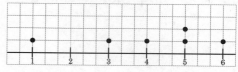
   (c)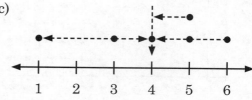

   (d) 1; 3
   3; 1
   4; 0
   5; 1
   5; 1
   6; 2
   Sum = 8

   (e) MAD: 1.33

4. (a) 33.5
   (b) 12; 21.5   21; 12.5
       39; 5.5   13; 20.5
       41; 7.5   36; 2.5
       49; 15.5   37; 3.5
       42; 8.5   45; 11.5
       MAD = 10.9 clicks

   (c)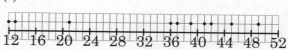

   (d) The mean is not a good indicator.

5. (a) Imani: 22
       Logan: 22

   (b) Imani's number of pages

       9 12 15 18 21 24 27 30 33 36 39 42

       Logan's number of pages

       9 12 15 18 21 24 27 30 33 36 39 42

   (c) Imani's distribution

   (d) Imani: 2
       Logan: 10.57

   (e) Answers will vary.

6. (a) Sacramento: 1.53
       San Francisco: 1.98
       Bakersfield: 0.54

   (b) Sacramento: 1.04
       San Francisco: 1.28
       Bakersfield: 0.41

   (c) Sacramento: LQ = 0.24;
       UQ = 2.97
       San Francisco: LQ = 0.18;
       UQ = 3.86
       Bakersfield: LQ = 0.08;
       UQ = 1.08

   (d) Sacramento: 2.73
       San Francisco: 3.68
       Bakersfield: 1

7. Sacramento:
   MIN = 0.04 in
   Lower quartile, Q_1 = 0.24 in
   Median, Q_2 = 1.04 in
   Upper quartile, Q_3 = 2.97 in
   MAX = 3.62 in

   San Francisco:
   MIN = 0.06 in
   Lower quartile, Q_1 = 0.18 in
   Median, Q_2 = 1.28 in
   Upper quartile, Q_3 = 3.86 in
   MAX = 4.57 in

   Bakersfield:
   MIN = 0 in
   Lower quartile, Q_1 = 0.08 in
   Median, Q_2 = 0.41 in
   Upper quartile, Q_3 = 1.08 in
   MAX = 1.22 in

8. (a) MIN = 0 min
       Lower quartile, Q_1 = 10 min
       Median, Q_2 = 23.5 min
       Upper quartile, Q_3 = 30 min
       MAX = 60 min

   (b)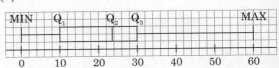

       Answers will vary.

9. Answers will vary. The sum of each data set must be 70.